Bridge Design, Assessment and Monitoring

This book serves as a valuable reference to all concerned with bridge design, assessment and monitoring, including students, researchers, engineers, consultants and contractors from all areas of bridge engineering. The contribution in each chapter presents state-of-the-art as well as emerging applications related to key aspects of bridge engineering.

The chapters in this book were originally published as a special issue of *Structure and Infrastructure Engineering*.

Airong Chen is Professor at the Department of Bridge Engineering, Tongji University, China. He is the Vice-Chair of Bridge and Structural Engineering Institute of China's Highway & Transportation Society. His main research areas involve bridge aerodynamics, bridge design theory, and life-cycle bridge engineering. He is one of the designers of the Sutong Bridge, a cable-stayed bridge with main span of 1088m. He received several national awards, including the First Class National Award for Science and Technology Progress, and the Senior Research Prize of the International Association for Bridge Maintenance and Safety (IABMAS).

Dan M. Frangopol is Professor and the inaugural holder of the Fazlur R. Khan Endowed Chair of Structural Engineering and Architecture at Lehigh University, USA. He has authored/co-authored over 350 articles in archival journals including 9 prize-winning papers. He is the Founding Editor of *Structure and Infrastructure Engineering* and of the book series *Structures and Infrastructures*. He is a foreign member of the Academia Europaea and of the Royal Academy of Belgium, an Honorary Member of the Romanian Academy and a Distinguished Member of the American Society of Civil Engineers (ASCE).

Xin Ruan is Associate Professor at the Department of Bridge Engineering, Tongji University, China. His main research areas involve bridge risk assessment and management, bridge loads, and bridge durability. He was awarded the Youth Science and Technology Award of China's Highway &Transportation Society in 2014.

Bridge Design, Assessment and Monitoring

Editor in Chief: Dan M. Frangopol

Edited by
Airong Chen, Dan M. Frangopol and Xin Ruan

LONDON AND NEW YORK

First published 2018
by Routledge
2 Park Square, Milton Park, Abingdon, Oxon, OX14 4RN, UK

and by Routledge
52 Vanderbilt Avenue, New York, NY 10017

First issued in paperback 2020

Routledge is an imprint of the Taylor & Francis Group, an informa business

© 2018 Taylor & Francis

British Library Cataloguing in Publication Data
A catalogue record for this book is available from the British Library

ISBN 13: 978-0-367-57220-4 (pbk)
ISBN 13: 978-0-8153-8228-7 (hbk)

Typeset in Minion Pro
by RefineCatch Limited, Bungay, Suffolk

Publisher's Note
The publisher accepts responsibility for any inconsistencies that may have arisen during the conversion of this book from journal articles to book chapters, namely the possible inclusion of journal terminology.

Disclaimer
Every effort has been made to contact copyright holders for their permission to reprint material in this book. The publishers would be grateful to hear from any copyright holder who is not here acknowledged and will undertake to rectify any errors or omissions in future editions of this book.

Contents

Citation Information

The chapters in this book were originally published in *Structure and Infrastructure Engineering*, volume 13, issue 4 (April 2017). When citing this material, please use the original page numbering for each article, as follows:

Chapter 9

Edgar Cardoso: a tribute to a brilliant bridge engineer
Paulo J. S. Cruz
Structure and Infrastructure Engineering, volume 13, issue 4 (April 2017), pp. 517–536

For any permission-related enquiries please visit:
http://www.tandfonline.com/page/help/permissions

Notes on Contributors

Mitsuyoshi Akiyama is Professor at the Department of Civil and Environmental Engineering, Waseda University, Japan.

Colin C. Caprani is Senior Lecturer at the Department of Civil Engineering, Monash University, Australia.

Joan R. Casas is Professor at the Department of Civil and Environmental Engineering, UPC-BarcelonaTech, Spain.

F. Necati Catbas is Professor at the Civil, Environmental and Construction Engineering Department, University of Central Florida, USA.

Airong Chen is Professor at the Department of Bridge Engineering, Tongji University, China.

Marios K. Chryssanthopoulos is Professor at the Department of Civil and Environmental Engineering, University of Surrey, UK.

Christian Cremona is Scientific and Technical Director at Bouygues Construction, France.

Paulo J. S. Cruz is Professor at the School of Architecture, University of Minho, Portugal.

Dan M. Frangopol is Professor and the inaugural holder of the Fazlur R. Khan Endowed Chair of Structural Engineering and Architecture at Lehigh University, USA.

Michel Ghosn is Professor at the Department of Civil Engineering, The City College of New York, USA.

Boulent Imam is Senior Lecturer at the Department of Civil & Environmental Engineering, University of Surrey, UK.

Alexandros N. Kallias is Former Research Fellow at the Department of Civil & Environmental Engineering, University of Surrey, UK.

Tung Khuc is Assistant Professor at the National University of Civil Engineering, Vietnam.

Rujin Ma is Associate Professor at the Department of Bridge Engineering, Tongji University, China.

Zichao Pan is Assistant Professor at the Department of Bridge Engineering, Tongji University, China.

Benoît Poulin is Senior Bridge Engineer at CEREMA in Nantes, France.

Xuefei Shi is Professor at the Department of Bridge Engineering, Tongji University, China.

Miriam Soriano is PhD student at the School of Civil Engineering, Technical University of Catalonia-BarcelonaTech, Spain.

Koshin Takenaka is Structural Engineer, Shimizu Co. Ltd., Japan.

Man-Chung Tang is Chairman of the Board TY Lin International San Francisco, USA.

Xin Ruan is Associate Professor at the Department of Bridge Engineering, Tongji University, China.

Junyong Zhou is PhD student at the Department of Bridge Engineering, Tongji University, China.

Bridge design, assessment and monitoring

The recent research progress in bridge maintenance, safety, management and life extension has been presented and discussed at the Seventh International Conference on Bridge Maintenance, Safety and Management (IABMAS 2014), held in Shanghai, China, July 7–11, 2014. The First (IABMAS'02), Second (IABMAS'04), Third (IABMAS'06), Fourth (IABMAS'08), Fifth (IABMAS 2010) and Sixth (IABMAS 2012) International Conferences on Bridge Maintenance, Safety and Management were held in Barcelona, Spain, July 14–17, 2002, Kyoto, Japan, October 18–22, 2004, Porto, Portugal, July 16–19, 2006, Seoul, Korea, July, 13–17, 2008, Philadelphia, PA, USA, July 11–15, 2010 and Stresa, Italy, July 8–12, 2012, respectively.

IABMAS 2014 (http://www.iabmas2014.org) has been organised on behalf of the International Association for Bridge Maintenance and Safety (IABMAS, http://www.iabmas.org) under the auspices of Tongji University. The objective of IABMAS is to promote international cooperation in the fields of bridge maintenance, safety, management, life-cycle performance and cost for the purpose of enhancing the welfare of society. The interest of the international bridge engineering community in the fields covered by IABMAS has been confirmed by the significant response to the IABMAS 2014 call for papers. In fact, over 600 abstracts from nearly 40 countries were received by the Conference Secretariat, and approximately 70% of them (i.e. 396 papers) were selected for final publication as technical papers and presented at the Conference within mini-symposia, special sessions and general sessions.

The extended versions of selected papers presented at IABMAS 2014 are included in this special issue of *Structure and Infrastructure Engineering*. Tang specifies the purposes and requirements of the conceptual bridge design, considering bridge types, basic elements, structural systems and load conditions. Cremona and Poulin propose an assessment procedure for existing bridges. Kallias et al. develop a framework for the performance assessment of metallic bridges under atmospheric exposure by integrating coating deterioration and corrosion modelling. Soriano et al. employ a simplified approach to estimate the maximum traffic load effect on a highway bridge and compare the results with other approaches based on on-site weigh-in-motion data. Akiyama et al. propose a method for reliability-based durability design and service life assessment of reinforced concrete deck slab of jetty structures. Chen et al. propose a meso-scale model to simulate the uniform and pitting corrosion of rebar in concrete and to obtain the crack patterns of the concrete with different rebar arrangements. Ruan et al. present a traffic load model for long span multi-pylon cable-stayed bridges. Khuc and Catbas implement a non-target vision-based method for the measurement of both static and dynamic displacements time histories. Finally, Cruz presents the career of the outstanding bridge engineer Edgar Cardoso in the fields of bridge design and experimental analysis.

The Guest Editors wish to thank the authors for their contributions as well as the reviewers for their timely assessment of the merit of the submitted manuscripts. It is expected that this issue will serve as a valuable reference to students, researchers, and will be helpful to all involved in bridge engineering.

Airong Chen

Dan M. Frangopol

Xin Ruan

Conceptual design of bridges

Man-Chung Tang

ABSTRACT

The design process of a bridge can be divided into four basic stages: conceptual design, preliminary design, detailed design and construction design. The purpose of the conceptual design is to come up with various feasible bridge schemes and to decide on one or more final concepts for further consideration. The purpose of the preliminary design is to select the best scheme from these proposed concepts and then to ascertain the feasibility of the selected concept and finally to refine its cost estimates. The purpose of the detailed design is to finalise all the details of the bridge structure so that the document is sufficient for tendering and construction. Finally, the purpose of the construction design is to provide step-by-step procedures for the building of the bridge.

Introduction

The design process of a bridge can be divided into four basic stages: conceptual design, preliminary design, detailed design and construction design (Tang, 2014). The purpose of the conceptual design is to come up with various feasible bridge schemes and to decide on one or more final concepts for further consideration. The purpose of the preliminary design is to select the best scheme from these proposed concepts and then to ascertain the feasibility of the selected concept and finally to refine its cost estimates. The purpose of the detailed design is to finalise all the details of the bridge structure so that the document is sufficient for tendering and construction. Finally, the purpose of the construction design is to provide step-by-step procedures for the building of the bridge. Each of the earlier design stages must carefully consider the requirements of subsequent stages. For example, the detailed design must consider how the bridge is to be built; the preliminary design must consider, in addition, how structural details will look like; and, the conceptual design must consider, in addition to all the above, what information the preliminary design will require. This means that a conceptual design must sufficiently consider what is required to complete the bridge in the given environment, including a general idea of costs and construction schedule as well as aesthetics. Schematically it can be represented by the sketch as shown in Figure 1, which shows that a conceptual design must be able to encompass the preliminary design, the preliminary design must be able to encompass the detailed design and the detailed design must be able to encompass the construction design.

By 'consider', we do not mean that we have to actually perform detailed studies on the aforementioned issues during the conceptual design stage, since accumulated engineering experience can help us understand the feasibility of many basic ideas. For example, we do not have to conduct a calculation to ascertain that a 150 m span prestressed concrete box girder bridge is feasible if we can assume the girder depth to be about 7.50 m deep. Rather, our experience accumulated from working on many other bridges tells us that, typically speaking, a medium span prestressed concrete box girder with a depth of 1/20th of the span length can work. By contrast, if we want to build a girder that is only 5.00 m deep, this *would* require a detailed study during the conceptual design stage because it is far less than the conventional girder depth of 1/20th of the span. Experience is the utmost of importance during the conceptual design stage. For this reason alone, only an experienced engineer should be appointed to perform a conceptual design.

Thus, a conceptual design can be defined as a process that must consider all details of the bridge in all phases from beginning to completion, at least based on experience if not in actual analysis. This is to ascertain that the proposed concept is feasible under the given conditions. Here, 'feasibility' should not be restricted merely to structural stability and constructability; it must also satisfy the four basic requirements of a bridge: safety, functionality, economy and aesthetics.

From another view point, the conceptual design stage initiates the design of a bridge. It is a 'conceiving' stage that begins with a blank slate – a creative process that tests the innovative abilities of the engineer.

The requirements of a good bridge

A successful bridge must satisfy four basic requirements: safety, functionality, economy and aesthetics. Safety cannot be

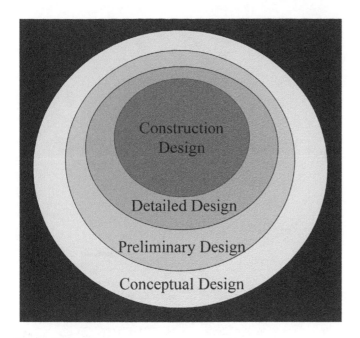

Figure 1. The four design stages.

compromised. A bridge must be safe under all of the loads it is designed for. Otherwise, the bridge can't be opened to traffic. And, when we talk about safety, we must make sure the bridge is safe not only for a certain limited period of time, but for the entire service life of the bridge. Thus, durability of the structure is equally important and it is contained in the consideration of safety. We have not satisfied the requirement of safety if a bridge designed for a service life of 100 years, for example, becomes unsafe after 50 years.

Functionality is the reason why we build a bridge in the first place. Functionality should not be compromised. If four lanes are required, for example, it must provide four lanes. But in certain cases, it may still be acceptable if some lane widths are slightly different from the standard width recommended in the specifications, as long as this does not affect safety.

Economy and aesthetics do not have absolute standards. There is no such thing as a 'correct cost' of a bridge. It varies from place to place, from time to time and from situation to situation. The cost of a bridge in Florida may be vastly different from the cost of that same bridge if built in New York, or in Shanghai, even though the structures are exactly the same. So what is economical is clearly a relative term. But still, if a bridge is to be built in a certain place, at a certain time and under a certain circumstance, there is a way to judge its economy. However, the first cost of a bridge is not the only cost item we have to consider, the maintenance expense is also a big item to be considered. In other words, the life cycle cost is the real cost of a bridge. If a bridge costs less to build but cost much more to maintain, it is not an economical bridge. Durability again plays a big role in this respect. In summary, economy can be expressed by a ratio of value vs. cost (Tang, 2010), R_{vc} = value/cost. The higher this ratio, the better the economy of the bridge.

Aesthetics are even harder to define. For example, the Firth of Forth Bridge in Scotland was often criticised as one of the world's most ugly structures by some and at the same time lauded as a spectacularly beautiful landmark by others. The Eifel Tower

in Paris was mercilessly attacked by many architects, engineers, philosophers and other intellectuals alike, as an eyesore of the City at the time of its construction. But now, it has become the most beloved tourist attraction in France. Nevertheless, people do admire the beauty of some bridges and dislike others. If a bridge is deemed beautiful by most people, then we may be allowed to say that it is beautiful. So, at least at a given time and in a given place, it is possible to say whether a bridge is beautiful or not, based on popular opinion.

Purpose of building a bridge

Before we embark on the process of conceptualising a bridge, we must understand why we build the bridge in the first place. The basic purpose of a bridge is to carry traffic over an opening or discontinuity in the landscape. Various types of bridge traffic can include pedestrians, vehicles, pipelines, water and ships, or a combination thereof. An opening can occur over a highway, a river, a valley or any other type of physical obstacle. The need to carry traffic over such an opening defines the function of a bridge. The design of a bridge can only commence after its function has been properly defined. Therefore, the process of building a bridge is not initiated by the bridge engineer. Just like roads or a drainage system, or other types of infrastructure, a bridge is a part of a transportation system and a transportation system is a component of a city's planning efforts or its area development plan. The function of a bridge must be defined in these master plans.

However, besides carrying traffic, a bridge may have to serve other functions as well, such as being a memorial to certain event, or as a signature monument of a city. These requirements could have significant influence on how the bridge should be conceptualised. In other words, the design of a bridge must duly consider what the stakeholder's expectations are and these considerations must be done in the conceptual design stage.

Basic bridge types

We can group all bridges in the world into four basic types: Girder bridge, arch bridge, cable-stayed bridge and suspension bridge, see Figure 2. There are also varying possible combinations, such as the cable-stayed and suspension scheme proposed by Franz Dishinger, see Figure 3, and the 'partially cable-supported girder bridge', see Figure 4. The partially cable-supported girder bridge is a combination of a girder bridge and any one of the aforementioned bridge types (Tang, 2007). The extradosed bridge is a special subset of the partially cable-supported girder bridge.

Common wisdom suggests that girder bridges and arch bridges are good for short to medium spans, while cable-stayed bridges are good for medium to longer spans, and suspension bridges are good for very long spans. Based on this assumption, some engineers established rules to assign a span range for each of these bridge types. For example, in the 1960s, the reasonable maximum span length of a cable-stayed bridge was thought to be around 450 m and that of a girder bridge was thought to be about 250 m. These previously held theories didn't last long as cable-stayed bridges with spans of over 1000 m have been completed since then.

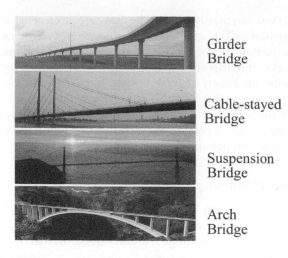

Girder
Bridge

Cable-stayed
Bridge

Suspension
Bridge

Arch
Bridge

Figure 2. The four basic types of bridges.

Figure 3. Dischinger combination.

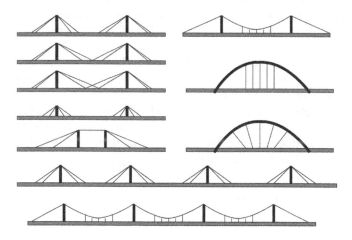

Figure 4. Partially cable-supported girder bridge.

Figure 5. Current world record span bridges (2014).

Over time, with improvements in construction materials and advancements in construction equipment and technique, the reasonable span length of each bridge type has significantly increased. But in relative terms, the above-mentioned comparisons are still valid. Only the numerical values of the span ranges have changed. However, an engineer should not be restricted by these assumptions. Instead, it is important for engineers to understand what the actual limits of each bridge type are, based on his or her understanding of the latest construction materials and equipment available at the time of construction. When a better material is available in the future, the engineer should be able to re-estimate these limitations using the same logic.

Nevertheless, it is important to understand what the maximum, technically feasible span length of each of these four types of bridges is. Based on the steel we have today, [tang], the maximum feasible span lengths are: 8000 m for a suspension bridge, 4200 m for an arch bridge, 5500 m for a cable-stayed bridge and about 550 m for a girder bridge. The world record spans are, correspondingly, the 1991 m span Akashi Kaikyo Suspension Bridge in Japan, the 1104 m span Russky Cable-Stayed Bridge in Russia, the 552 m span Chaotianmen Arch Bridge in China and the 330 m span Shibanpo Girder Bridge in China, see Figure 5. They are well under the maximum technically feasible span lengths. This shows that in our process of composing a new bridge, the maximum span length is not a restriction. In other words, in technical sense, we do not have to worry about the span of our bridge being impossible to build. However, 'feasible' is not the same as 'appropriate'. When the span gets longer, the cost of the bridge will go up very fast. Therefore, we should not increase the bridge span beyond what is necessary. It will unnecessarily increase the cost.

Basic elements of a bridge structure

Irrespective of how a structure looks like, all bridge structures are comprised of four basic structural elements, and each one is

The " A-B-C " of basic structural Elements

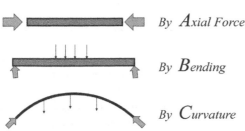

By Axial Force

By Bending

By Curvature

Figure 6. Structural elements A, B and C.

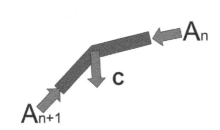

Figure 7. Formation of a curved element.

Figure 8. Stress distribution in an A element.

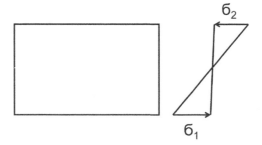

Figure 9. Stress distribution in a B element.

dominated by one type of function. They are: axial force elements (A elements), bending elements (B elements), curved elements (C elements) and torsional elements (T elements), which can be abbreviated as the ABCT of structures. Actually, the first three types of elements are sufficient to compose almost all structure types, see Figure 6. Most torsional elements can be established using a combination of the first three element types. But, for convenience, having torsional elements will simplify our thinking process.

As an example, in a cable-stayed bridge, the predominant function of the cables, the girder and the towers is to carry axial forces. These are mainly A elements. The same is true for a truss bridge. Certainly, there are local effects that may cause bending moment in these elements, but they are less dominant and can be considered secondary. A girder bridge, on the other hand, carries the loads mainly by bending, so it is considered a B element.

When an axial force element changes direction, it creates a force component lateral to the axial force, see Figure 7. So each change of direction will create a lateral component. These lateral components can be used to resist lateral loads. If the lateral loads are sufficiently closely spaced, the structural element becomes a curve, resulting in a curved structural element, the C element. There are two major types of curved elements: if the axial force is compression, the structure is similar to an arch. If the axial force is tension, the element is similar to a suspended cable such as the main cable of a suspension bridge.

Thus, with A, B and C elements, we can create the framework of almost all major types of bridges known to us today. Torsion is usually a locally occurring phenomenon. It mainly co-exists with one of the A, B or C elements. An eccentrically loaded girder bridge, for example, will have torsion besides bending moment. While both of them must be considered in the design, the predominant factor is still the bending moment and it therefore can be characterised as a B element.

In a design, we proportion the structural elements to remain within the allowable stress limits. Figures 8 and 9 show the stress distribution of the A element and the B element. In an A element, the entire cross-section can be utilised to its fullest extent because the entire cross-section can reach the allowable stress at the same time. By contrast, in a B element, only the extreme fibre can reach the allowable stress, while the stress in the rest of the cross-section is less than the allowable stress. So in a B element, most of the cross-sectional area is not fully utilised and is consequently less efficient. The C element is similar to the A element and it is more efficient than the B element.

When a portion of the element is not participating in carrying loads, or if it is not used to its fullest extent, more material is required to carry the same load, thus increasing its dead weight, which is a big disadvantage in bridges, especially in long-span bridges. Currently, the world record span for each of these four types of bridges are the following: the 330 m span Shibanpo Bridge in Chongqing, China is the longest girder bridge span; the 552 m span Chaotianmen Bridge, also in Chongqing, China is the longest arch bridge span; the 1104 m Russky Bridge in Russia is the longest cable-stayed bridge span; and the 1991 m span Akashi Kaikyo Bridge in Japan is the longest suspension bridge span, see Figure 5. The Shibanpo Bridge, a girder bridge, over the Yangtze River in Chongqing is the smallest among world record spans of the four bridge types. As explained above, the girder bridge is the only bridge type that relies mainly on bending to carry the loads. It is a B element. B elements are less efficient. Therefore, a bridge that consists mainly of a B element is less efficient and thus its maximum span is smaller. Keeping this in mind, we can proceed to compose the type of structure we like for a bridge under a set of given conditions.

Subsets of building blocks

We can now go one step further to establish a few basic subsets of building blocks which can be employed in the composition of a bridge structure:

- *The Triangle:* A triangle is comprised of three axial force elements. A triangle is a very stable structure by itself. The triangle is the most commonly used subset in bridge structures. A truss bridge is nothing but a group of triangles. A cable-stayed bridge is also a composite form of triangles.
- *The Tied-Arch:* An arch works only if the horizontal force in both ends of the arch is resisted by some means. When an arch rib is loaded only by vertical loads, which is the case for almost all bridges, the horizontal reaction at both ends of the arch is equal to each other but in opposite directions. A tie can be used to tie the arch ends together so there will be no horizontal reaction required of the foundations.

Feasible structural systems

A bridge is usually analysed as a two-dimensional structure. Thus, it has to satisfy three basic equations according to applied mechanics:

$$\left(\Sigma F_x = 0; \Sigma F_y = 0; \Sigma M = 0. \right)$$

Any structure that can fully satisfy these three equations is a feasible structure. Any structure that cannot fully satisfy these three equations is not structurally feasible. This feasibility is not limited to the entire structure, it is also required for any free-body cut out from the structure.

Obviously, a bridge is not a two-dimensional structure even though it is often analysed as two-dimensional structural system for simplicity purposes which is usually sufficient for global design. If a three-dimensional analysis is necessary, we will have six equations to deal with instead of the three basic equations of mechanics above:

$$\left(\Sigma F_x = 0; \Sigma F_y = 0; \Sigma F_z = 0; \Sigma M_x = 0; \Sigma M_y = 0; \Sigma M_z = 0 \right)$$

Again, any structure that can fully satisfy these six equations is a feasible structure. Any structure that cannot fully satisfy these six equations is not structurally feasible. Hence, when we start composing a structure, examine it thoroughly to ascertain that it can satisfy these equations. If not, the bridge cannot be built.

Permanent load condition

As noted above, when we analyse a bridge as a two-dimensional structure, it has only to satisfy three basic equations according to applied mechanics:

$$\left(\Sigma F_x = 0; \Sigma F_y = 0; \Sigma M = 0. \right)$$

Any structure that can fully satisfy these three equations is a stable structure. This seemingly simple criterion is a very powerful tool in a conceptual design. While it is not possible to alter the stress distribution in a statically determinate structure, there are numerous possible patterns of stress distributions in a statically indeterminate structure. This can be illustrated by a simple example of a two-span bridge, see Figure 10. If the two spans are not continuous, the bridge consists of two statically determinate, simply supported beams. If it is continuous, it is a one degree statically indeterminate structure, which means there is one unknown when we try to calculate the forces and bending moments in the structure using the three equations above. This allows us to assign any value to the unknown and still satisfy these three equations, thus creating a stable structure.

For simplicity's sake, let us assume the bending moment at the middle support, M_b, is the unknown. If we set M_2 to zero, the bending moment in the bridge will be the same as two simply supported beams. However, we can assign any other value to M_b, and for each value we assume, we will get a different bending moment diagram in the bridge. Thus, by varying the value of M_b, we can optimise the efficiency of the bridge girder. There are many ways to accomplish the desired value of M_2 in actual construction. One simple method is to adjust the reaction at the centre support. This can be done by using hydraulic jacks. By adjusting this reaction, any value of M_b can be achieved.

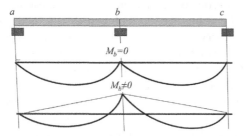

Figure 10. A two-span beam.

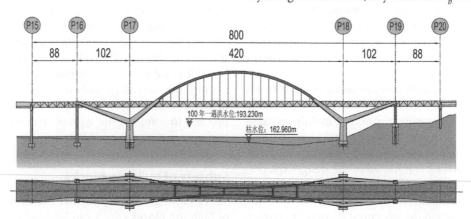

Figure 11. Caiyuanba Bridge, Chongqing, China.

Figure 12. Concrete-filled steel tube arch bridge.

A three-span continuous girder is a two degree statically indeterminate structure with two unknowns, the value of which we can assume as we prefer. Accordingly, an N degree statically indeterminate structure will have N unknowns, the values of which we can also assume. Thus, for the permanent load condition, we should be able to calculate the stress distribution of the entire bridge simply using the three basic equations mentioned above. This should be a rather simple calculation even for a highly redundant structure such as a cable-stayed bridge with many cables. While the calculation is simple, the selection of the most appropriate permanent load condition is not. In the final design, the bridge must also sustain many additional loads such as live loads, wind loads, earthquake loads, ship impact, etc. Therefore, in the process of assigning a permanent load condition to the bridge, we must also consider all load combinations that the bridge must endure.

Examples

Following are two examples of how a bridge concept is developed based on the 'tools' that were described above. In the process,

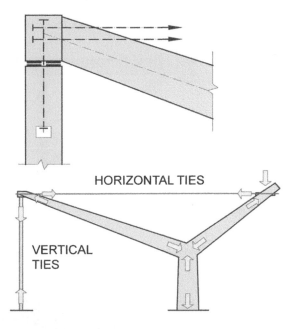

Figure 14. Details of the concrete Y-frame.

common sense is very important. Very often, how we feel about a structural system may tell us a lot about the appropriateness of our concept.

Caiyuanba Bridge, Chongqing, China

The City of Chongqing is rather hilly. The landscape is very graceful. I used to compare Beijing as a seriously looking Mandarin, Shanghai as a talkative merchant, Wuhan as a muscular worker and Chongqing as a beautiful lady. A bridge in such a landscape as Chongqing's should be as slender and graceful as possible. This led to my selection of the basket handle, tied-arch bridge, with a uniform box section for the arch ribs, see Figure 11.

The water level of the Yangtze River at this location may vary up to about 38 m, which requires special considerations for possible barge collision and corrosion protection. An arch with its

Figure 13. Comparison of two schemes.

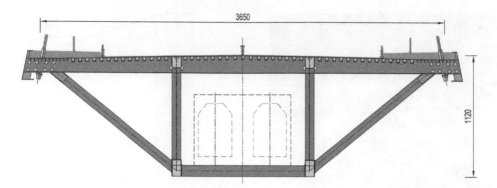

Figure 15. Girder cross-section.

Figure 16. Appearance of truss girder.

Figure 17. Erection of arch ribs.

arch ribs partially submerged in water would not be aesthetically pleasing! So, I raised the arch high above the normal water level to rest on four vertical concrete columns; in that way, it is above the highest level of water. I used concrete for the lower portion of the bridge because concrete would fare better in case of a possible barge collision; concrete would also be more resistant when submerged in flood waters.

Aesthetically, the concrete frame makes the bridge appear sturdier. China has built a large number of concrete-filled

steel-tube arches, see Figure 12. This type arch was also seriously considered as an alternative to the box-shape arch ribs. However, a concrete-filled steel-tube arch would look too bulky at this location, see Figure 13. Conceptually, under permanent loads, the structural system can be treated as a simple, tied-arch bridge resting on two concrete 'Y' frames. All three ties can be individually stressed to achieve the optimal internal force distribution, see Figure 14.

Each tie is comprised of several groups of individually sheathed and epoxy-filled seven wire strands. Each strand can be replaced individually, if necessary. Each concrete 'Y' frame has two ties; a horizontal tie and a vertical tie-down at the end. The global bending moment can be eliminated by properly adjusting the forces in these two ties. The girder is about 12 m deep to provide sufficient room to accommodate the monorail on the lower level, see Figure 15. Typically this type of girder would look very bulky over the river. I chose a truss to make the girder look more transparent, see Figure 16. A truss also offers the monorail passengers a more open view of the river.

In addition, the trapezoidal-shaped cross-section makes the girder look much more slender.

The in-plane buckling of an arch bridge depends on the combined stiffness of the girder and of the arch rib acting together. Hence, the arch ribs of this bridge can be made extremely slender because the girder is very stiff. Chongqing is also called the 'fog city of China'. It is foggy most of the time. As all polluting factories had been relocated to outside, all buses and taxis are now running exclusively on natural gas, and most residents have switched from using coal to gas, the fog situation has improved significantly. But the city is located inside the Yangtze and Jialing River Valleys, the moisture cannot easily escape. Fog is still very common. In foggy weather, visibility is important. As steel bridge must be painted, selection of the appropriate colour of the paint is an important part of the design. Yellow and orange-red are the two most visible colours in fog and these were the two colours we proposed for the arch ribs. These two colours also match very well with the grey colour of the concrete frames. The Owner selected the orange-red colour.

The construction scheme is a part of a bridge design. One of the major considerations in construction is the supply of materials. Chongqing is a hilly city with narrow and winding streets. It is basically impossible to deliver construction materials and equipment through these streets in the city. Therefore, river transportation is the only logical means of delivery. But the

Figure 18. Erection of girder.

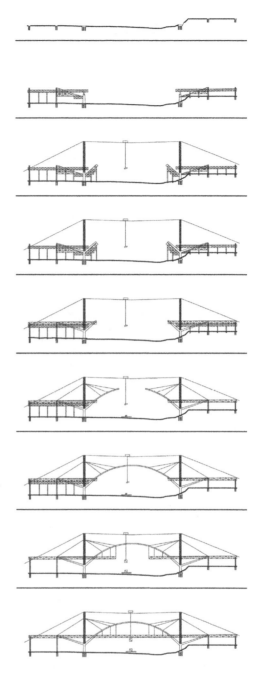

Figure 19. Erection sequence.

large variation of water levels in Chongqing's rivers is always a major problem for bridge construction. The normal high water level and the normal low water level of the Yangtze River at this site differ by about 23 m. The highest high water can rise to 38 m above the lowest low water level. Besides, during the low water season, only the southern portion of the river is navigable which makes docking difficult.

To overcome this difficulty, I recommended a construction scheme using high lines. The fact that the area had good rock footing for anchoring the back-stays of the high line was important in the selection of this concept. The contractor erected the world's heaviest high line with a capacity of 420 tons. Actually, after modifying the lifting procedure, the heaviest piece to be lifted was only 350 tons.

Construction started by building the prestressed concrete Y frame on local falsework. The arch ribs were erected using the high lines segmentally from both ends simultaneously and symmetrically, at a speed of about 3-day cycles, see Figure 17. The arch ribs were held in place by a temporary stay cable system. After completion of the arches, the truss girder was also erected using the high lines, see Figure 18. The closure operation went very smoothly. It is important that the temporary stays must be adjustable to correct any possible geometry deviation during erection. This way the closure operation will be very simple, see Figure 19.

This example described how a conceptual design is done. We started with a set of given conditions:

- owner's desire of an arch bridge;
- a main span length of 420 m;
- high variation of water level and possible ship impacts;
- double deck girder which usually looks bulky;
- transportation difficulties in material supply;
- graceful landscape but foggy; and
- owner's desire of a good-looking bridge.

During the conceptual design stage, we considered all these given conditions and came up with a concept that solved all these problems, not only design problems, but also the method of construction. The method of construction is very important because restriction at the site may render a bridge concept in-appropriate due to cost or time schedule. And, all along we kept both cost and aesthetics in mind. It is important to note that in the process of the entire conceptual design, all decisions were based on experience or some very simple hand calculations. There was no time for lengthy analysis.

Jiayue Bridge, Chongqing, China

The Jiayue Bridge, see Figure 20, crosses the Jialing River near the town of Yuelai, Chongqing. The surrounding area is to be developed into a high-end residential neighborhood. Therefore, the Owner desired to build a beautiful bridge as an attraction to potential homeowners.

The bridge deck is about 100 m above the water level in the river connecting the top of the river valley at both ends. Thus, the bridge superstructure is located above the land at both sides, see Figure 21. For such a landscape, a traditional cable-stayed

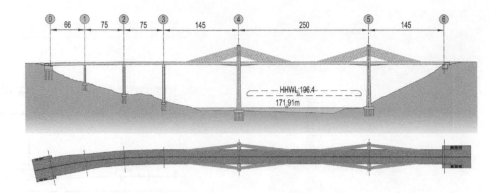

Figure 20. Jiayue Bridge over the Jialing River, Chongqing, China.

Figure 21. The superstructure is above the surrounding landscape.

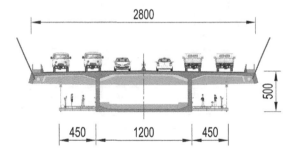

Figure 22. Cross section of girder.

Figure 23. Pedestrian path.

bridge was not ideal because the towers would have been too tall and it would have had too many cables that could look quite distracting. Hence, a partially cable-supported girder bridge that would have shorter towers is a better configuration. However, the towers should not be too low so as to become inefficient.

The river traffic here consists mainly of barges and fishing boats. The required main span was 250 m. However, the bridge has a total length of about 800 m. The upper portions of the towers are made to lean outward to give the passengers on the bridge a very open view. With the concrete technology today, such sloping columns are just as easy to build as a vertical one. The bending moment caused by the inclination of the tower columns can be compensated by the eccentricity of the cables at the girder level.

The Jiayue Bridge carries 6 lanes of highway traffic and 2 pedestrian/bicycle paths. The bridge deck would be very wide if we put them all in one level. This region of China is called one of the three 'ovens' of China, indicating how hot it can be in the summer. Temperatures can be over 40 °C for prolonged periods

during the summer and it is rather rainy in the winter. To offer pedestrians refuge from the sun and rain, the pedestrian paths were placed underneath the wing slabs of the upper deck, see Figures 22 and 23. This also resulted in a narrower bridge deck, which is structurally more efficient.

When we put the pedestrian paths under the deck, pedestrians may feel boxed in when the walkway is right next to a wall. Therefore, such a pedestrian path must be sufficiently wide to offer more openness to make the pedestrians more comfortable. The walkways are suspended from the cross beams of the deck on the outside and are attached to the web wall on the inside. The walkway deck is made of steel to reduce the overall weight.

For security reasons, we did not provide any crosswalk between the two walkways for the entire length of the bridge.

However, there is a wide, open space underneath each end span so that people can walk from one side to the other side. There is also room for other facilities if necessary. The walkway deck is covered with a colored mat to offer a more soothing feeling underfoot. The walls may be used for art exhibitions during special occasions. All these decisions were made during the conceptual design stage. Again, there was no need to have any detailed structural analysis for the structure. The entire conceptual design was accomplished using simple hand calculations.

Summary

Conceptual design is where bridges are conceived. A good conceptual design must properly consider what may be required in the preliminary design stage, detailed design stage and construction design stage; it must assure that the proposed concept is safe, functional, economical and pleasing to the eye. Because every bridge is unique, the conceptual design must begin by being unique. The conceptual design is our best opportunity for innovative thinking and it is where new ideas are conceived!

A bridge is built for the public; we must be conservative in the design to assure safety and functionality. But during conceptualisation, we must strive to be as innovative as possible. Only by pushing the boundaries of what we perceive as possible can we extend the horizons of engineering and thus improve our civilisation. 'Conservative Innovation' is the path to follow in the conceptual design of a bridge!

Disclosure statement

No potential conflict of interest was reported by the author.

References

Tang, M. C. (2007, July). First person: Rethinking bridge design: A new configuration. *Civil Engineering Magazine Archive, 77,* 38–45.

Tang, M. C. (2010, October 27–30). *Cost versus value. Proceedings of the 2nd Symposium on Life-Cycle Civil Engineering.* International Association for Life-Cycle Civil Engineering, Taipei, Taiwan.

Tang, M. C. (2014). Conceptual design. In W.-F. Chen & L. Duan (Eds.), *Bridge engineering handbook, Second Edition: Fundamentals* (pp. 1–28). Boca Raton, FL: CRC Press, Taylor and Francis Group.

Standard and advanced practices in the assessment of existing bridges

Christian Cremona and Benoît Poulin

ABSTRACT

In recent years, the condition of deficient bridges has reached such a level that the volume of required repair actions becomes significant for many countries. As budgets for maintenance, repair and rehabilitation are always limited and demands are constantly increasing, to find an optimal balance between cost and safety is today a new trend in bridge maintenance. Optimising bridge maintenance and management is a strong expectation for owners and stakeholders facing ageing bridge stocks and increasing aggressive traffic. In this context, the assessment of the structural performance may be necessary for various reasons thorough its lifetime. In France, there are no standards or regulations for structural assessment of existing structures. The studies on a new Eurocode standard for the 'Evaluation and rehabilitation of existing structures' are just starting and it will be published in several years. For this reason, the French Ministry of Transport has decided to develop recommendations for the assessment of the structural performance of existing bridges. In a first part, the paper summarises today's practice in France, but it also details the ongoing calibration process for setting appropriate partial factors for existing bridges. Implementation is given as example for the assessment of reinforced concrete slabs.

Introduction

Optimising bridge management is a challenge for owners and operators who are faced with ageing structures and increased aggressive traffic loads. Within this context, the performance assessment of a structure may be necessary during its lifetime for various reasons, such as

- to take account of changes in operating conditions,
- to assess its ability to carry exceptional convoys without induced damage,
- to evaluate its load-carrying capacity in cases of damage and to conclude whether or not an intervention (repair, load limit, etc.) is mandatory.

Performance assessment is therefore not restricted to deteriorated structures, and a large proportion of this activity is also related to bridges in good condition just to check whether they are able to carry exceptional loads. Several countries have already edited guidelines on this subject (AASHTO, 2011; HA, 2001; DB, 1999; SIA, 2009; CSA, 2006 etc.). The French technical corpus is conversely limited and Eurocodes explicitly exclude existing structures from their scope, even if CEN (2002, 2004b) mention that they can be used for the assessment of existing structures (with amended provisions) to plan repairs or to study operating conditions. It is important to note that the rules set down in design codes constitute a set of prescribed rules that are only valid within a certain context. Thus, in order to be applicable, a bridge must conform to the design code in the following areas:

- the type of bridge,
- the methods used in the structural analysis,
- the quality of construction materials and workmanship,
- the actual traffic loading on the bridge,
- the condition of the bridge,
- the detailing used.

For assessment, situations often exist which render design codes inapplicable either because of existing structural condition or because of the presence of non-conforming details. This is particularly in the case of older bridges, and current design codes have to be interpreted carefully before being used. A new Eurocode on the 'Assessment of existing structures' is currently on progress but will be published in several years.

Design codes present safety margins which, in general, exceed those that are reasonable to accept for the assessment of existing bridges. This is because the level of knowledge related to the materials, loading conditions and structural behaviour can be identified with a greater degree of reliability with *in situ* measurements. Theoretically, partial factors (PFs) can be modified while maintaining the same level of structural reliability.

It is clear that the establishment of principles and procedures to be used for the assessment of existing bridges is needed because some aspects of assessment are based on an approach that is substantially different from new design, and requires knowledge beyond the scope of design codes. In addition, bridge assessment should be carried out in stages of increasing sophistication, aiming at greater precision at each higher level. In order to save structures from unnecessary rehabilitation or replacement (and therefore to reduce owners' expenditure), the engineer must use all the techniques, all the methods and all the information available in an efficient way. Simple analysis can be cost-effective if it demonstrates that the bridge is satisfactory, but if it does not, it can present major drawbacks regarding the bridge under study and advanced methods should be introduced as an option as far as possible.

Due to the lack of doctrine in France, upon the request of the Directorate for Transportation Infrastructures (French Ministry of Ecology, Sustainable Development and Energy in charge of transport and mobility regulations), the Technical Department of Transportation Infrastructures and Materials of the CEREMA institute has initiated the development of recommendations for the assessment of the structural performance of existing structures. This project covers two actions: the first task is to clarify, unify and formalise the practices of the different offices of the Directorate for Transportation Infrastructures (DIT). The general principles of this state of the art are briefly recalled in the first part of this paper. The second task is to cover the research on the calibration of PFs fitted on investigation results. This constitutes the second and major part of the present paper. This second part provides a general approach for PF updating and is not restricted to the French or European context.

The novelty of this paper is the introduction of adapted code calibration procedures for existing structures when *in situ* data are available, as an alternative to generic equations (such as given in Equations (3) and (4)). Indeed, most of the authors today (see for instance Koteš & Vičan, 2013; Sýkora, Holický, & Marková, 2013) use this approach in connection with target reliability indexes as given in Eurocodes. In this paper, a long section will be dedicated to emphasise the limitations of this approach and its inconsistency for non-linear limit states with more than three variables. In order to bypass the problem, code calibration procedures have been used in the paper, in addition to the calculation of target reliability indexes based on statistical assumptions. This complexifies the calculations but provides a coherent framework for the calibration of the PFs. At last, the proposed approach avoids the numerical complexity of the usual code calibration procedures and introduces several improvements to provide efficient sets of PFs.

Assessment levels

The French practice in structural assessment is initially carried out using simple methods but more refined methods are used if the required capacity is higher than the assessed capacity. Four assessment levels are proposed (Sétra, 2012a).

- Level 0 concerns structures that are accepted into the management system without a formal assessment. Documents and inspection records must have been analysed and the structural conditions do not raise concerns.
- Level 1 requires carrying out a formal assessment based on design standards. This level is only prescribed for undamaged bridges.
- Level 2 makes use of tests or investigations to obtain bridge-specific data. This level implies adjustments of material properties and loadings, and eventually structural

behaviour. For Level 2, semi-probabilistic methods are used but revised partial factors are allowed to account for bridge-specific information. These partial safety modifications must be based on probabilistic reliability assessments and specific tabulated data.

Level 3 is a full reliability analysis. Important matters to settle are model uncertainties, the target reliability and factors affecting these quantities. These assessments usually require specialist knowledge and expertise and are likely to be worthwhile only in exceptional cases.

Performance assessment

The idea that an existing structure should not be evaluated by standards calibrated for new structures is generally accepted. However, even if the regulations for new structures are not enforced, it remains necessary to rely on regulations/standards to carry out the calculations. In doing that, opportunities should be sought to relax the design rules based on more up-to-date information and research. It is advantageous to use the same format for the assessment rules as used in the corresponding design standard. Mistakes can be reduced and the structural engineer can identify differences more easily.

Selection of a design standard

In the absence of assessment guidelines, it is important to refer to design rules. There are two main options: those given by the standard used for the design of the bridge of those currently applied. For a long time, this question was not fully addressed. The French DIT recommends (Sétra, 2012a) using the current regulations, i.e. Eurocodes, since they are considered as the most advanced ones, reflecting the current scientific knowledge. However, in the case of structural assessments under exceptional loads, it remains usual to merely compare the effect of the exceptional loading to the effect of the design loads (i.e. using the regulations in force at the construction time), and to consider this checking as acceptable if the exceptional load effect is less aggressive. This approach is only applicable to structures designed with so-called modern rules, i.e. with safety levels close to those prescribed by Eurocodes (that is to say, structures built after 1965).

Selection of a loading standard

The principle adopted in the French recommendations (Sétra, 2012a) is to use the current loading standards, i.e. CEN (2003), that is assumed supposed to be representative of European traffics. It should be noted that this regulation has been calibrated on the basis of measured traffic on motorways carrying heavy traffics. It is well-suited for structures supporting very heavy traffics. For others, when verification is not satisfied with CEN (2003), it is possible to perform the structural assessment in accordance with the former bridge loading code (MEL, 1971). The results must be analysed with respect to the comparisons between the two approaches and the traffic level effectively carried out by the bridge.

In general, CEN (2003) and MEL(1971) are relatively homogeneous in terms of safety level. However, due to a larger load eccentricity imposed by CEN (2003), this latter loading code can be significantly more aggressive for certain structures (such as girder bridges). Frequent load values (lower than the characteristic values) can be chosen for short-time duration works. It may also be possible to use values issued from traffic measurements: the characteristic values are then extrapolated from measurements data (Cremona, 2001). This approach can reduce traffic load effects for a given structure. However, such an approach requires validation by a technical committee as there is a risk to underestimate the load level.

Studies, based on traffic measurements of very long durations (one year of continuous monitoring), are under progress by the Technical Department of Transportation Infrastructures and Materials of the CEREMA to adjust the coefficients of the Uniform Distributed Load and Truck Systems loads given in CEN (2003). These coefficients are indeed supposed to cover all types of structures and may be excessive for some of them. Such an approach has been introduced in Switzerland and led to significant reductions in load PFs (OFROU, 2006).

Limit states

According to CEN (2002), serviceability limit states considers

- the functioning of the structure or its structural components under normal operation,
- the users' comfort,
- the construction aesthetics.

The principle that an existing structure can have a serviceability level lower than the one required for a new structure is usually applied by the French DIT. Nevertheless, any modification of the serviceability level must rely on technical studies and economic considerations and is forbidden for structural assessments under exceptional convoys.

CEN (2002) states that ultimate limit states (ULSs) are related to the safety of the users and of the structure itself. With regard to human safety, the DIT enforces that an existing structure must have a safety level similar to those required for new structures. This principle is consistent with the 'strength and stability' CEN (2002) requirements that must fulfil construction works. Although modifications of the safety level are forbidden, it is nevertheless possible to allow amendments (such as prescribed by CEN (2002)):

- Modification of PFs for ULS,
- introduction of measured values of loading and/or material properties.

Modification of the PFs

The ISO (2010) states in its paragraph 7.3 that the PFs, for new structures, can be modified to take into account the results from inspections and tests. The design codes use safety margins that, in general, exceed those that are reasonable to accept for the assessment of existing bridges. This is because the level of knowledge of existing structures and/or the actual traffic conditions can be determined to a greater degree of reliability, as they can be observed and/or measured. Thus, PFs can be reduced while maintaining the same level of reliability. Therefore, knowledge

regarding strength and loading can be increased by further investigations and this can justify modifying PFs. Reducing uncertainty about some variables justifies a reduction of the related PFs without reducing the safety level.

Investigations improving knowledge about fixed and dead loads are for instance:

- Measurements of pavement thickness and super-structures,
- geometry measurements (beams, slabs, etc.),
- concrete density measurements from samples,
- weighing of the support load effects …

Similar approach can be applied for material properties (yielding or ultimate strength…). The uncertainties in the measurement of these quantities must be provided. As previously mentioned, traffic loads can be optionally modified by adjusting traffic coefficients or by taking account of Weigh-In-Motion data.

PF updating

Principles

Bridge assessment is very similar to bridge design. The same basic principles lie at the heart of the process. An important difference nevertheless lies in the fact that when a bridge is being designed, an element of conservatism is generally a good thing that can be achieved with very little additional costs. When a bridge is being assessed, it is important to avoid unnecessarily conservative measures because of financial implications that may follow if a bridge is designated as substandard without good cause.

The *Structural Reliability Theory*, which expresses structural safety in probabilistic terms, can be an adequate answer for bridge assessment. However, such an approach induces some difficulties, theoretical and numerical as well as practical. In turn, the *semi-probabilistic approach* as used in many design codes or standards (Cremona,2011) schematically replaces this probability calculation by the verification of a criterion involving design values of strength R_i and load S_j, noted $R_{d,i}$ and $S_{d,j}$:

$$\mathcal{G}\left(R_{d,i}, S_{d,j}\right) \geq 0 \leftrightarrow \mathcal{G}\left(\frac{R_{k,i}}{\gamma_{R_i}}, \gamma_{S_j} S_{k,j}\right) \geq 0 \qquad (1)$$

in which $\mathcal{G}(.)$ represents the performance (or limit state) function: negative values of this function correspond to the failure domain, while positive ones are related to the safety domain. $\mathcal{G}(.) = 0$ defines the limit state. $R_{k,i}$, $S_{k,j}$ correspond to nominal values associated with strength and load variables. $\gamma_{R_i}, \gamma_{S_j}$ are the related partial (safety) factors.

The partial safety approach is claimed to be semi-probabilistic, considering the application of statistics and probability in the evaluation of the input data, the formulation of assessment criteria and the determination of load and resistance factors. The semi-probabilistic code format is therefore deterministic, but the PFs are the solution of reliability and optimisation procedures. These PFs also cover uncertainties induced by parameters omitted during the study and are fitted to provide designs not too different from designs made by former design methods. Their essential vocation is to cover the variability in structural behaviour and loading.

The use of reliability (probabilistic) techniques for structural assessment is not common in structural engineering. Yet, it can be very valuable because such approaches can handle uncertainties in a more efficient manner than a semi-probabilistic as introduced in codes or standards. In particular, semi-probabilistic formats are unable to efficiently utilise information obtained from measurements and inspections. To overcome this limitation, a probabilistic approach can be used to re-calibrate PFs.

EN1990 and ISO2394

The CEN (2002) and the ISO (1998) standards provide mathematical expressions for calculating PFs for resistance properties, as the ratio of design and characteristic values:

$$\gamma_R = \frac{R_k}{R_d} \qquad (2)$$

Table C3 of CEN (2002) lists formulas for the design values depending on the probability model used for describing the variables (lognormal, normal, Gumbel...). For instance, for a lognormal distribution, it comes, respectively, for the design and the characteristic values (5% fractile):

$$R_d = \frac{\mathbb{E}[R]}{\sqrt{1 + \mathbb{C}\mathbb{o}\mathbb{V}[R]}} \exp\left(-\alpha_R \, \beta_t \, \sqrt{\ln\left(1 + \mathbb{C}\mathbb{o}\mathbb{V}[R]\right)}\right) \qquad (3)$$

$$R_k = \frac{\mathbb{E}[R]}{\sqrt{1 + \mathbb{C}\mathbb{o}\mathbb{V}[R]}} \exp\left(-1.645 \, \sqrt{\ln\left(1 + \mathbb{C}\mathbb{o}\mathbb{V}[R]\right)}\right) \qquad (4)$$

where $\mathbb{E}[R]$ is the mean value of variable R, $\mathbb{C}\mathbb{o}\mathbb{V}[R]$ is the coefficient of variation, β_t is the target reliability level and α_R is the sensitivity factor.

Theoretically the sensitivity factors are determined from reliability analyses. They are direction cosines of the normal to the limit state function transformed in a space of standardise normal variables. In this space, the point on the limit state that is the closest to the origin is the design point. CEN (2002) and ISO (1998) propose values for a resistance PF, i.e. .8 for a dominating resistance parameter and .32 for the other (non-dominating) resistance parameters.

Similar approaches stand for loading variables and the direction cosines of .7 are often prescribed:

$$\frac{1}{\gamma_S} = \frac{S_k}{S_d} \qquad (5)$$

Intuitively, these expressions can be used for modifying PFs as soon as a new mean value and a new coefficient of variation are available from tests. Several papers have been recently published using that approach (Koteš & Vičan, 2013; Sýkora et al., 2013). Equations (2)–(4) ensure the target reliability level β_t. Nevertheless, Equations (2)–(4) present serious drawbacks that have to be pointed:

- These expressions are well-suited for linear limit state functions with two/three variables. For more than two

variables, the direction cosines can hardly fulfil the condition $\|\{\alpha\}\| = \sqrt{\sum \alpha_i^2} = 1$; as it will be shown later, the proposed values (.8 or .7) are rarely obtained for complex limit state functions;

- these equations assume that the direction cosines are not modified by the change in the probabilistic parameters: this is not true in reality. To be convinced, let us note that for a linear limit state with two normal variables, the α_R coefficient is equal to $\sigma_R \big/ \sqrt{\sigma_R^2 + \sigma_S^2}$. Changing the standard deviation of a variable will affect the direction cosines;

- it is well known that there exists an infinity of PFs for a given set of variables in a limit state. Using Equations (2)–(4) for PF updating can lead to a strong deviation from the initial set of PFs;

- these expressions do not take into account correlations between variables;

- some concern can be raised about the choice of the target reliability level. CEN (2002) or ISO (1998) are providing values. As it will be highlighted later, these values strongly differ from what usual designs can provide. Consequently, they have to be used very carefully.

All these aspects explain that Equations (2)–(4) have not been adopted in the following studies. To do so, it is necessary to revisit the code calibration process. The purpose of this article was therefore to introduce a code calibration process based on minimisation procedure instead to use generic mathematical expressions that are too simplistic.

Code calibration

The objective of code calibration is to provide PFs for the design of a class of structures. The general procedure is based on the determination of the 'best' set of PFs, which leads to structures as close as possible to the code objective expressed in terms of allowable target reliability. The procedure is divided into five main steps (Gayton, Mohamed, Sorensen, Pendola, & Lemaire, 2004):

- Step 1: definition of the class of structures (failure mode(s) and the limits of validity of the code). This step requires the definition of mechanical and probabilistic models.

- Step 2: definition of the target reliability level β_t within the structure class of Step 1. The European standard CEN (2002) provides target reliability indexes for ultimate and serviceability limit states. These values reflect a sort of European consensus for minimal performance level that can be very far from usual designs and strongly depends on the choices fixed for describing the probabilistic nature of the variables. For this reason, the CEN (2002) target reliability index will not be used in the present study but instead minimum performance designs (i.e. corresponding to $\mathcal{G}(.) = 0$) will be calculated according to the design situations given in Step 3 and reliability analyses will be carried out to assess the corresponding reliability indexes.

- Step 3: selection of L design situations in the structure class. The design situations should cover the whole set of cases that may occur during design.

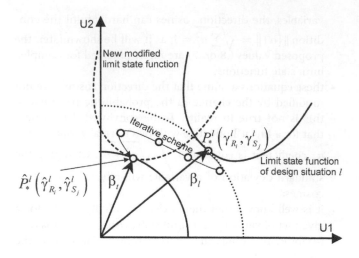

Figure 1. FDP procedure.

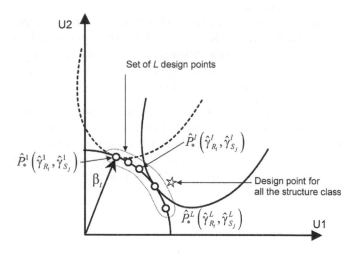

Figure 2. PF computation.

- Step 4: target fitting. The choice of a penalty function $\mathcal{W}\left(\beta_l\left(\gamma_{R_i},\gamma_{S_j}\right),\beta_t\right)$ is chosen for a specific design situation. This penalty functions depends on the set of PFs. Usually, it is expressed in terms of reliability index $\beta_l\left(\gamma_{R_i},\gamma_{S_j}\right)$ than probability of failure since the first indicator is usually calculated before the probability of failure (as it is in FORM/SORM methods).

- Step 5: optimisation of the PFs. This is performed by solving the following optimisation problem where $\gamma_{R_i},\gamma_{S_j}$ are the optimisation variables:

$$\min_{\gamma_{R_i},\gamma_{S_j}} \sum_{l=1}^{L} \mathcal{W}\left(\beta_l\left(\gamma_{R_i},\gamma_{S_j}\right),\beta_t\right) \qquad (6)$$

(L being the number of design situations). The solution of this minimization problem is not unique and there exists an infinity of PFs verifying the minimal condition (Cremona, 2011).

- Step 6: partial factors checking. Random situations are generated within the structure class, and reliability indexes are evaluated for minimum performance design $\mathcal{G}(.) = 0$. The agreement between the obtained reliability index and the target index β_t indicates the quality of the calibration.

Step 5 is the most difficult part to perform. Two different approaches can be used for solving Equation (6):

- Global optimisation procedure (Kroon, 1994) that consists in directly solving Equation (6). This procedure is time-consuming and requires a large number of reliability analyses;

- approximate methods (Ditlevsen & Madsen,1996) that are based on a decomposition of step 5 into two sequential substeps. This procedure is less time-consuming but gives a lower accuracy than the global optimisation procedure.

Approximate calibration methods

To decrease the number of reliability analyses required by the global optimisation method, approximate methods have been introduced. Several approaches are available and descriptions, analyses and comparisons can be found in Gayton et al. (2004).

Approximate methods are divided into two steps. The first step is called *Fitted Design Point* (FDP) search while the second one is the *PF* computation. In the FDP search, a design situation '*l*' and an arbitrary set of PFs $\left(\gamma_{R_i}^l,\gamma_{S_j}^l\right)$ are chosen. In the normalised U-space, an iterative scheme (Figure 1 shows the process in a plan combining two normalised variables U1 and U2) allows the modification of the limit state function in order to reach the code objective (target reliability index β_t), i.e. starting from design point $P_*^l\left(\gamma_{R_i}^l,\gamma_{S_j}^l\right)$ to reach new design point $\hat{P}_*^l\left(\hat{\gamma}_{R_i}^l,\hat{\gamma}_{S_j}^l\right)$ that belongs to the hyper-sphere of radius β_t. For each single design situation, a set of PFs $\left(\hat{\gamma}_{R_i}^l,\hat{\gamma}_{S_j}^l\right)$ is deduced from the $\hat{P}_*^l\left(\hat{\gamma}_{R_i}^l,\hat{\gamma}_{S_j}^l\right)$ since the PFs are the control variables in this iterative scheme.

There are several FDP methods, the most common one being to translate the design points $P_*^l\left(\gamma_{R_i}^l,\gamma_{S_j}^l\right)$ to reach the distance β_t. This method needs only one single reliability analysis but has the drawback to propose a design point not corresponding to the most likelihood failure point. Let us note that this is exactly the approach used for establishing designs values of Table C3 from CEN (2002). In this paper, the global optimisation problem as expressed by Equation (6) is replaced by L individual optimisation calculations:

$$\min_{\gamma_{R_i}^l,\gamma_{S_j}^l} \mathcal{W}\left(\beta_l\left(\gamma_{R_i}^l,\gamma_{S_j}^l\right),\beta_t\right) \qquad (7)$$

This presents the advantage to reduce the computational complexity, but requires to find the 'best' PFs that will be valid for the whole structure class, once a set of PFs $\left\{\left(\hat{\gamma}_{R_i}^l,\hat{\gamma}_{S_j}^l\right)\right\}_{1\leq l\leq L}$ is obtained. This is the purpose of the second substep called partial factor (PF) computation. This consists in the computation of the 'best' PFs on the basis of the L sets of PFs obtained in FDP substep. It is generally based on the definition of a single design point from the L design situations; the chosen design point should be representative for the whole possible designs in the structure class (Figure 2).

Several methods are also available and Gayton et al. (2004) details and compares some of them. The simplest one is to take:

$$\tilde{\gamma}_{R_i} = \max\left(\hat{\gamma}_{R_i}\right); \tilde{\gamma}_{S_j} = \max\left(\hat{\gamma}_{S_j}\right) \qquad (8)$$

This choice leads to a conservative design point: by construction, its reliability index is larger than the target reliability index. To avoid the overrepresentation of any deviant design situation, it is instead proposed to fix the final set of PFs $\left(\tilde{\gamma}_{R_i}, \tilde{\gamma}_{S_j}\right)$ as a 95% fractile estimation, assuming a Gaussian distribution:

$$\tilde{\gamma}_{R_i(S_j)} = \mathbb{E}\left[\hat{\gamma}_{R_i(S_j)}\right] + 1.65\, \sigma\left[\hat{\gamma}_{R_i(S_j)}\right] \tag{9}$$

where $\mathbb{E}[.]$, $\sigma[.]$ denote the mean and standard deviation of a random variable. Equation (9) will be used for determining the set of PFs for the structure class. This equation avoids including too specific design situations and helps to maintain a certain coherence among the final set of PFs. It is nevertheless necessary to check that the final set of PFs does not provide deviant results in terms of reliability level for these specific design situations.

Selection of a target reliability index (step 2)

The selection of the target reliability index is important. Three choices can be made:

- To use one of the target reliability indexes as given in CEN (2002) (preferably coming from the reliability class RC2),
- To choose a reliability index, representing all the design situations and based on the individual allowable indexes given for each design situation (i.e. $\mathcal{G}(.) = 0$)
- To keep all the individual allowable reliability indexes from each design situation.

The first option has been previously discussed and excluded. The second option leads to use a target reliability index that can be higher or lower compared to the different design situations and introduce an additional variability in the final set of PFs. The third option consists in substituting β_t (in Equation (7)) by the individual allowable reliability indexes $\beta_{t,l}$. This option eliminates the variability induced by a single target reliability index. In summary, Equation (7) is modified into:

$$\min_{\gamma_{R_i}^l, \gamma_{S_j}^l} \mathcal{W}\left(\beta_l\left(\gamma_{R_i}^l, \gamma_{S_j}^l\right), \beta_{t,l}\right) \tag{10}$$

The use of Equation (10) is only possible because each design situation has a specific target reliability index. This target reliability index corresponds to the minimum allowable design.

Selection of coefficients to update (step 5)

Not all PFs deserve to be updated. This is only meaningful for PFs associated with variables that may be measured during inspections and investigations, or are the most sensitive for the reliability analysis. Let us note that some PFs are implicitly fixed to 1.00 and are therefore hidden in the calculation process. Consequently, if data are available for the related variable, it may be necessary to introduce such a PF for which the initial value in the design regulation is equal to 1.00.

A measured quantity X will be characterised by a bias v_X and its coefficient of variation (due to uncertainty measurement) $\mathbb{C}o\mathbb{V}_X$:

- $v_X = \frac{\bar{X}}{X_k}$ ($\bar{X}$ is the average value from *in situ* tests and X_k the nominal/characteristic value),

- $\mathbb{C}o\mathbb{V}_X = \frac{\sigma_X}{\bar{X}}$ (σ_X is the standard deviation of the measurement X).

Checking updated PFs (step 6)

The major drawback of the PF computation is that mixing different variables for a specific design situation cannot be necessary in agreement with the allowable reliability index $\beta_{t,l}$. Said differently, the final set of PFs $\{\tilde{\gamma}\}$ as given by Equation (9) can provide an allowable reliability index $\tilde{\beta}_{t,l}$ that may differ from the allowable reliability index $\beta_{t,l}$ for each design situation. The first reason is that the set $\{\tilde{\gamma}\}$ is composed of fractiles, and it is therefore necessary how far from the deviant situations it is. The second reason is that the PFs are individually updated, and it is necessary to assess the variation of $\beta_{t,l}$ with respect to $\beta_{t,l'}$

Application of reinforced concrete slabs

Step 1: structure class

The approach is applied to the reliability of reinforced concrete slabs (Figure 3). These bridges usually have rectangular cross sections although other designs are available. The slabs are cast *in situ* and consist of 2, 3 or 4 spans.

Limit state function

Before to calculate the target reliability index, it is necessary to recall the basic design principles of CEN (2005). Structural assessment consists of balancing resisting and bending characteristic moments by determining a minimum reinforcement area. The concrete mechanical behaviour is described by a simplified rectangular law as given in Figure 4, with:

- Strength and deformation coefficients, ηf_{cd} and ε_{cu}:

$$\varepsilon_{cu} = 0.35\%$$
$$\lambda = 0.8 \text{ and } \eta = 1.0 \text{ for } f_{ck} \leq 50\,\text{MPa} \tag{11}$$

- design compressive concrete strength f_{cd}:

$$f_{cd} = \alpha_{cc}\frac{f_{ck}}{\gamma_c} \text{ with } \alpha_{cc} = 1.0 \text{ and } \gamma_c = 1.5 \tag{12}$$

where f_{ck} is the concrete characteristic strength, and α_{cc} is a coefficient taking into account long-term effects of the material.

The constitutive law for reinforcement steels is an elastic-plastic diagram with no ultimate elongation. The design yield strength is given by:

$$f_{yd} = \frac{f_{yk}}{\gamma_s} \text{ with } \gamma_s = 1.15 \tag{13}$$

The minimum reinforcement area $A_{s,0}$ is corresponding to a *strict* balance between resisting and bending moments. The different stages for this calculation are the following:

- Moment equilibrium

$$\begin{aligned}\mathcal{M}_{\text{load},k} &= \mathcal{M}_{\text{strength},k} \\ &= b\,\lambda\,\eta f_{cd}\left(d - \frac{\lambda x}{2}\right) \\ &= 0.8\,\alpha(1 - 0.4\,\alpha)b\,d^2 f_{cd}\end{aligned} \tag{14}$$

17

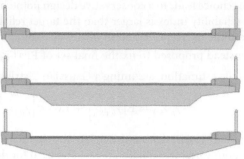

Figure 3. Reinforced concrete slab bridges: general view and typical cross sections.

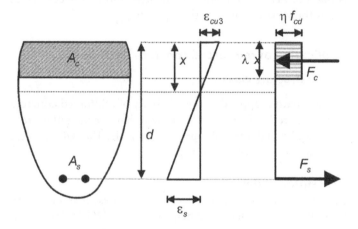

Figure 4. Equilibrium of a concrete cross section and simplified strain–strength concrete diagram.

where $\alpha = x \, d$ is the relative neutral axis of the cross section. b, d are the width and the distance of the reinforcement from the slab top, respectively (rectangular cross section).

- Calculation of the reduced moment μ:

$$\mu = \frac{M_{\text{load},k}}{b \, d^2 \, f_{cd}} = 0.8 \, \alpha (1 - 0.4 \, \alpha) \qquad (15)$$

- Calculation of the neutral axis:

$$\alpha = 1.25 \left(1 - \sqrt{(1 - 2 \, \mu)} \right) \qquad (16)$$

- Calculation of the axial force and of the steel deformation:

$$\mathcal{N}_{\text{load},k} = 0.8 \, \alpha \, b \, d \, f_{cd}$$
$$\varepsilon_s = \frac{1 - \alpha}{\alpha} 0.35\% \qquad (17)$$

- Calculation of the reinforcement area depending on the functioning domain (elastic or plastic):

$$\varepsilon_s < \frac{f_{yd}}{E_s} : A_{S,0} = \frac{\mathcal{N}_{\text{load},k}}{\varepsilon_s E_s}$$
$$\varepsilon_s \geq \frac{f_{yd}}{E_s} : A_{S,0} = \frac{\mathcal{N}_{\text{load},k}}{f_{yd}} \qquad (18)$$

E_s is the steel Young's modulus.

Calculation of the minimum performance design

The minimum performance design consists in calculating the reinforcement area that balances the bending and resisting characteristic moments:

$$\mathcal{M}_{\text{strength},k} = \mathcal{M}_{\text{load},k} \qquad (19)$$

For the studied cross section, three bending moments must be identified:

- $\mathcal{M}_{D,k}$ the dead load bending moment due to self-weight,
- $\mathcal{M}_{S,k}$ the dead load bending moment due to fixed loads,
- $\mathcal{M}_{L,k}$ the dead load bending moment due to live loads.

As an example, let us consider the bridge OA1 given in Table 2. The ULS bending moment is therefore given by:

$$\begin{aligned} \mathcal{M}_{\text{load},k} &= \gamma_L \, \mathcal{M}_{L,k} + \gamma_S \, \mathcal{M}_{S,k} + \gamma_D \, \mathcal{M}_{D,k} \\ &= 1.35 \times 3.225 + 1.35 \times 0.665 + 1.35 \times 1.595 \quad (20) \\ &\approx 7.404 \text{ MN.m} \end{aligned}$$

where 1.35 is the PF for the ULS design. The resisting moment is expressed in terms of the reinforcement area A_s that depends on the slab cross section geometry and on the material properties:

- the concrete compressive strength $f_{ck} = 35$ MPa divided by the PF $\gamma_c = 1.5$,
- the yielding limit strength of the reinforcement bars $f_{yk} = 500$ MPa divided by the PF $\gamma_c = 1.15$.

Applying CEN (2005) leads to the minimal reinforcement area $A_{S,0} \approx 48.87$ cm^2 (i.e. corresponding to $\mathcal{G}(.) = 0$). The minimal reinforcement area is calculated at critical cross sections (hot points) where the bending moments reach the maximum values.

Reliability analysis

For the reliability analysis, the limit state function is derived from the design requirements (for a design situation l):

$$\mathcal{G}^l = A_s^l - A_{s,0}^l \qquad (21)$$

The resisting moment is depending on the cross section geometry, the bar characteristics (area, yielding stress, positioning) and the concrete properties (compressive strength).

To perform a reliability analysis, it is necessary to probabilistically model the various variables introduced in the limit state function. Table 1 presents the set of variables that have been

Table 1. Random variables used for the reliability analysis.

Variable	Nominal value	Probability distribution	Bias	Coefficient of variation (%)	Standard deviation
Reinforcement area A_s (m^2/m)	Tab. 2	Normal	1.00	2	–
Yield strength f_y (MPa)	500	Lognormal	1.15	5	–
Height uncertainty Δh (mm)	0	Normal	1.00	–	10 mm
Concrete cover uncertainty Δc (mm)	0	Normal	1.00	–	5 mm
Fixed load moment $\mathcal{M}_s$ (MN.m)	Tab. 2	Normal	1.00	20	–
Concrete density ρ_c (kN/m^3)	25	Normal	1.00	5	–
Compressive strength f_c (MPa)	Tab. 2	Lognormal	1.20	10	–

Table 2. Characteristics of the investigated bridges.

Bridge	OA1	OA2	OA4	OA6	OA45	OA44	OA42
Spans (m)	9.4	8.1	9.5	8.09	10.5	9.19	12
	15	12	15.21	12.02	14.33	12.28	
	9.4	12	15.21	12.03	10.17	9.19	
		8.1		9.5	8.09		
Skew (gradian)	100	100	98.14	91.88	100	70	108.7
Width (m)	10.77	10	11.2	8.12	8.56	8.56	9.6
Height (m)	.55	.49	.597	.5	.516	.446	.6
Concrete (MPa) – f_{ck}	35	35	30	30	32	32	30
Reinforcement (MPa) – f_y	500	500	500	500	500	500	500
Critical cross section(MS: mid-span/S: support)	MS 2	MS 2/3	MS 2/3	MS 2/3	MS 2	MS 2	MS 1
Fixed loads bending moment (MN.m)	0665	0334	0550	0298	.468	.325	2.426
Dead load bending moment (MN.m)	1595	0768	1588	0599	1051	0681	2580
Live load bending moment (MN.m)	3225	1995	2726	1845	2386	2006	4328
Minimal reinforcement area (10^{-4} m^2)	48.87	32.64	34.44	31.23	41.99	35.51	66.31

Bridge	OA41	OA40	OA39	OA38	OA37	OA36	OA35
Spans (m)	10.65	20	9.2	9	8.37	12.09	11.958
	14.00	18.41	13.5	14.50	11.17	16.15	17.27
	10.00			11.3	11.1	8.37	12.09	12.214
Skew (gradian)	78.44	100	100	86.3	85	70	70
Width (m)	6.62	11.81	6.61	6.61	8.96	8.96	8.96
Height (m)	.6	.69	.55	.55	.414	.564	.665
Concrete (MPa) – f_{ck}	32	35	32	32	32	32	32
Reinforcement (MPa) – f_y	500	500	500	500	500	500	500
Critical cross section(MS: mid-span/S: support)	MS 2	.4 S 1	MS 2	MS 2	MS 2	MS 2	MS 2
Superstructure moment (MN.m)	.331	1.062	.295	.368	.291	.646	.604
Dead load bending moment (MN.m)	0868	5878	0719	0896	0527	1466	2048
Live load bending moment (MN.m)	1634	6530	1569	1688	1845	2827	3028
Minimal reinforcement area (10^{-4} m^2)	31.99	81.84	31.61	36.59	33.53	46.38	45.36

identified as the most influencing ones. This selection has been based on the values of some sensitivity factors. Among them, the direction cosines of the limit state function at the design point provide the relative importance of the individual random variables for the reliability index. In addition, relative sensitivity measures (Cremona, 2011) express the sensitivity of the reliability index vs. the statistical parameters of the relevant variables (mainly mean and standard deviation). The probability functions have been based on the literature review (Cremona, 2011). The live load effects have been conserved as deterministic and equal to their characteristic values. For the concrete cover, the XC4 exposure concrete class and a project lifetime of 100 years have been assumed. Since the concrete strength belongs to the C35/45

class, under this assumption of 100 years, the required concrete cover category is S5 from CEN (2005), leading to a minimal cover of 35 mm. Adding a 5 mm execution tolerance, the nominal concrete cover is fixed to 40 mm.

Computation of the allowable reliability index

The allowable reliability index for a specific design situation is calculated with $A_s^l = A_{s,0}^l$:

$$\beta_{t,l} = -\Phi^{-1}\left(\mathcal{P}\left(A_{s,0}^l < A_s^l\right)\right) = -\Phi^{-1}\left(\mathcal{P}\left(\mathcal{G}^l < 0\right)\right) \quad (22)$$

Table 3. Grids of 'reinforcement area' PFs for reinforced concrete slabs (in italics: the set of PFs from CEN, 2005).

	Bias		2%	3%	6%
Reinforcement area	.80		1.00	1.05	1.63
		Max	1.00	1.05	1.68
				OA40	OA40
	.90		1.00	1.05	1.63
		Max	1.00	1.05	1.68
				OA40	OA40
	1.00		*1.00*	1.05	1.63

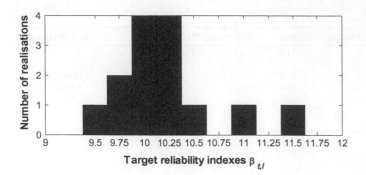

Figure 5. Distribution of the 14 target reliability indexes.

Steps 2 and 3: design situations and allowable reliability indexes

Fourteen design situations (slabs) have been considered in the present study (Table 2). They have been all designed according to CEN (2004a, 2005). The minimal reinforcement area has been calculated for the ULS and is given in Table 2. The selection of ULS is explained by the fact that for existing structures, this priority is given to structural safety compared to serviceability. These bridges have been calculated according to the French Sétra's guidelines (Sétra, 2012b).

Figure 5 presents the distribution of the target reliability indexes given by the different designs. A rather narrow variation of high values can be noticed. Target reliability indexes are very high compared to the values found ISO (1998) or CEN (2002). It is important to clarify this point. Firstly, ISO (1998) and CEN (2002) target reliability indexes are 'zero floor' values: nothing prevents to have larger values in design. Secondly, the

values calculated from the 14 situations describe the current design practice of reinforced concrete slab in France, and some over-dimensioned aspects may occur. Thirdly, all the reliability calculations are strongly depending on the choice of distribution functions describing the probabilistic nature of the variable. At last, the consistency between CEN (2002) target reliability indexes and PFs of the different Eurocodes is not straightforward. All these aspects make important to recalculate the target reliability indexes in order to have a consistent set of PFs for all the design situations.

Steps 4 and 5: PF updating

The PFU (PF updating) grids will consider:

- The PF γ_S for the fixed load moments, with a bias variation ranging from .80 to 1.40 (new bridge: 1.00), a coefficient of variation ranging from 10 to 20% (new bridge: 20%);
- the PF γ_A for the steel reinforcement with a bias variation ranging from .80 to 1.20 (new bridge: 1.00), and a coefficient of variation ranging from 2, 5, 8, 10 (new bridge: 2%);
- the PF γ_D for the concrete density, with a bias variation ranging from .90 to 1.10 (new bridge: 1.00), and a coefficient of variation ranging from 2, 5, 7, 10 (new bridge: 5%).

The approach introduced herein is to update the individual PFs vs. a range of bias and CoV values. This will lead to four tables, one for each variable: when more than one variable is modified, PFs from the two individual tables are combined. In the present paper, the open source RelsciTbx R1.0 toolbox (Sétra, 2011) (developed under Matlab© environment) has been used for determining the updated PFs.

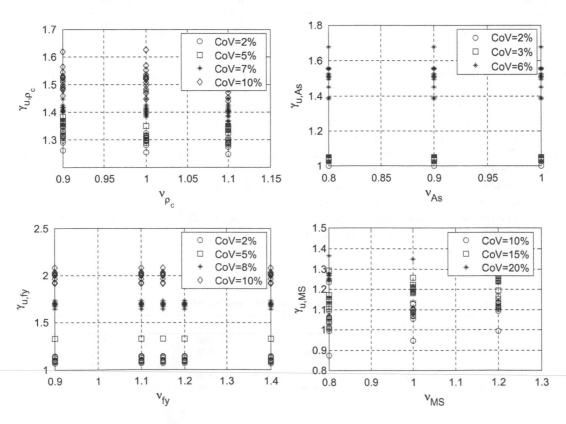

Figure 6. Variations of the PFs among the 14 designed slabs.

Table 4. Grids of 'yield strength' PFs for reinforced concrete slabs (in italics: the set of PFs from CEN, 2005).

	Bias		2%	5%	8%	10%
Yield strength	.900		1.13	1.32	1.73	2.07
		Max	1.14	1.32	1.73	2.08
			OA37	OA37	OA40	OA40
	1.100		1.13	1.32	1.73	2.07
		Max	1.14	1.32	1.73	2.08
			OA37	OA37	OA40	OA40
	1.150		1.13	*1.32*	1.73	2.07
		Max	1.14	1.32	1.73	2.08
			OA37		OA40	OA40
	1.200		1.13	1.32	1.73	2.07
		Max	1.14	1.32	1.73	2.08
			OA37	OA37	OA40	OA40
	1.400		1.13	1.32	1.73	2.07
		Max	1.14	1.32	1.73	2.08
			OA37	OA37	OA40	OA40

Table 5. Grids of 'concrete density' PFs for reinforced concrete slabs (in italics: the set of PFs from CEN, 2005).

	Bias		2%	5%	7%	10%
Concrete density	.900		1.35	1.37	1.43	1.58
		Max	1.34	1.38	1.45	1.62
			OA42	OA42	OA40	OA40
	1.000		1.33	*1.35*	1.43	1.59
		Max	1.33	1.35	1.45	1.62
			OA42		OA40	OA40
	1.100		1.31	1.35	1.44	1.60
		Max	1.30	1.35	1.45	1.63
				OA40	OA40	OA40

Table 6. Grids of 'fixed load moment' PFs for reinforced concrete slabs (in italics: the set of PFs from CEN, 2005).

	Bias		10%	15%	20%
Fixed load moment	.800		1.17	1.23	1.32
		Max	1.24	1.29	1.37
			OA40	OA40	OA40
	1.000		1.17	1.24	*1.35*
		Max	1.19	1.26	1.35
			OA40	OA40	
	1.200		1.18	1.27	1.45
		Max	1.16	1.26	1.43

Figure 6 depicts the scarcity of the obtained PFs for all the situations of the structure class. It can be noticed that the PFs for the resistance variables f_y and A_s are not depending on the bias. Conversely, the load variables ρ_c and $\mathcal{M}_S$ are bias-dependent. Tables 3–6 give the PFs obtained for all the four variables and for the entire family (14 design situations). It can be noticed that the PFs for the resistance variables f_y and A_s are not depending on the bias although this is not the case for load variables ρ_c and $\mathcal{M}_S$.

Tables 3–6 also include the maximum value obtained and the corresponding case of the structure class. OA37, OA40 and OA42 are design situations for which the proposed set of PFs is lower than the ones calculated for each bridge. For this reason, a special check is necessary. Let also note that some maximal values may be lower than the set of PFs for the structure class. This is due to the fact that the PFs are fractile values and nothing prevents to exceed the maximal values.

Two checks are performed for assessing the pertinence of the set of updated PFs for the structure class. The first one consists in calculating the minimal and the maximal values of the relative errors between the allowable reliability $\tilde{\beta}_{t,l}$ for the family set of PFs and the initial allowable reliability index $\beta_{t,l}$ related to the same design situations. This calculation is done for each

design situation and each variable. It helps to identify the design situations that have the largest negative errors (lower reliability index $\tilde{\beta}_{t,l}$): bridges OA37 and OA40. Two other design situations (OA4 and OA42) have smaller negative errors and are selected as well as four situations with no negative errors (OA1, OA39, OA2, OA41).

A second check is applied to these eight selected design situations. Allowable reliability indexes $\tilde{\beta}_{t,l}$ are recalculated when combining 2, 3 and 4 updated PFs. Negative and positive relative errors are again assessed. Only, negative errors are sensitive since they lead to lower reliability indexes $\tilde{\beta}_{t,l}$ compared to initial reliability indexes $\beta_{t,l}$.

Table 7 summarises these average different errors (when combining 2, 3 and 4 PFs) for the 8 selected design situations. The most unfavourable error is given for OA1 with a negative error of −2.47% when combining 2 PFs: let us recall that this design situation highlighted no negative error with one coefficient only. The average positive errors remain reduced with a maximum deviation of 12.88% for OA42 (combining 4 factors). For OA40, the average negative error does not exceed −2.20% (4 factors) but a large part (50%) of the reliability indexes $\tilde{\beta}_{t,l}$ are lower than $\beta_{t,l}$.

Table 7. Negative and positive relative errors from initial target reliability indexes.

	Relative error (%) between $\tilde{\beta}_{t,l}$ and $\beta_{t,l'}$							
	1 factor		2 factors		3 factors		4 factors	
Design situation	Neg.	Pos.	Neg.	Pos.	Neg.	Pos.	Neg.	Pos.
OA01	0	1.90	−2.47	3.59	−2.19	5.29	−1.92	7.00
Percentage of results	.00	100.00	1.29	98.71	3.75	96.25	7.95	92.05
OA40	−.60	3.65	−.95	5.22	−1.42	5.70	−2.20	6.25
Percentage of results	53.19	46.81	46.12	53.88	36.51	63.49	30.76	69.24
OA37	−1.44	2.86	−1.01	5.40	−.78	7.77	−.65	10.51
Percentage of results	10.64	89.36	10.34	89.66	6.19	93.81	2.93	97.07
OA39	0	2.22	−.39	4.23	−.48	6.12	−.50	7.93
Percentage of results	.00	100.00	1.29	98.71	2.63	97.37	3.86	96.14
OA02	0	2.11	.00	3.99	−.21	5.86	−.23	7.76
Percentage of results	.00	100.00	.13	99.87	1.50	98.50	2.78	97.22
OA04	~.00	2.07	−2.16	3.79	−2.06	5.44	−1.79	7.11
Percentage of results	2.13	97.87	1.42	98.58	3.75	96.25	7.87	92.13
OA41	0	2.25	−.79	4.26	−.84	6.13	−.72	7.94
Percentage of results	.00	100.00	1.42	98.58	3.04	96.96	5.63	94.37
OA42	−0.16	3.83	−0.17	7.20	−0.25	10.17	−0.33	12.88
Percentage of results	12.77	87.23	6.85	93.15	2.18	97.82	0.62	99.38

From all these results, it can be stated that the relative errors (especially the negative ones) are relatively small and that the set of updated PFs for the structure class can be considered as providing the same reliability level than the Eurocode standards. This validates the proposed approach consisting in the calibration of individual PF grids.

Application for the assessment of an existing structure

Let us consider that *in situ* measurements help to improve the knowledge about the bias $v_{\mathcal{M}_S}$ of the fixed load moment:

$$v_{\mathcal{M}_S} = \mathcal{M}_S^{\text{test}} / \mathcal{M}_{S,k} \qquad (23)$$

The nominal value $\mathcal{M}_{S,k}$ is the value prescribed by the design rule. In addition, a new coefficient of variation $\mathbb{CoV}_{\mathcal{M}_S}^{\text{test}}$ is determined during the *in situ* tests (variability of results within the sampling set). The PF updating procedure is intended to modify the initial PF γ_S in order to cope with the new information available from *in situ* measurements. If $\tilde{\gamma}_S$ is the updated PF for the fixed load moment, then the assessment rule (relatively to $\mathcal{M}_S$) will be written as it follows:

$$\hat{M}_{\text{load}} = \gamma_L\, M_{L,k} + \tilde{\gamma}_S\, M_S^{\text{test}} + \gamma_D\, M_{D,k} \qquad (24)$$

The design value $\mathcal{M}_{S,k}$ is therefore replaced by the mean test value $\mathcal{M}_S^{\text{test}}$. The same procedure applies for the other variables (concrete density, reinforcement area, yield strength).

Conclusions and perspectives

This paper presents the current approach for structural assessment and the ongoing studies. A procedure for calibrating PFs based on *in situ* experiments has been presented. Relying on FDP search and *Partial factor* (PSF) computations, it consists of updating PFs for each individual variable. For reinforced concrete slab bridges, four variables have been highlighted as significant and four tables have been given for different bias and coefficients of variation. When data are available at least for two variables, the PFs from these tables are combined.

This strategy is particularly interesting because it leads to introducing only one table per variable, although, in theory, combinations must be analysed. In turn, it is essential to check that the reliability indexes when crossing the individual tables remain equal or larger than the initial target reliabilities. This verification is essential for validating the calibration procedure that is proposed in this paper. Verification highlights that some case studies may introduce combinations that may induce allowable reliability indexes no more than 2.5% smaller than the target reliability indexes. The same procedure is currently implemented for the assessment of prestressed and reinforced concrete girders and slabs.

Acknowledgements

This research has been conducted at the Technical Centre for Bridge Engineering, Technical Department of Transportation Infrastructures and Materials, CEREMA, upon the request of the Directorate for Transportation Infrastructures of the Ministry of Transport. Its support is greatly appreciated. Authors are also thankful to C. Marcotte, J. Michel, R. Sadone and B. Vion from Cerema for their valuable contribution to this study and to A-S. Colas and A. Orcesi from Ifsttar for their advices.

Disclosure statement

No potential conflict of interest was reported by the authors.

References

AASHTO. (2011). *Manual for bridge evaluation* (2nd ed.). Washington, DC: American Association of State Highway and Transportation Officials.

CSA. (2006). *CAN/CSA-S61: Canadian highway bridge design code*. Rexdale, ON: Canadian Standards Association.

Cremona, C. (2001). Optimal extrapolation of traffic load effects. *Structural Safety, 23*, 31–46.

Cremona, C. (2011). *Structural performance: Probability-based assessment*. London: ISTE-Wiley.

DB. (1999). *DS 805: Tragsicherheitsnachweis bestehender Eisenbahnbrücken* [DS 805: Load carrying capacity assessment of existing rail bridges]. Hauptverwaltung der Deutschen Bundesbahn, Berlin, Germany. [in German].

Ditlevsen, O., & Madsen, H.O. (1996). *Structural reliability methods*. Chichester: Wiley.

CEN. (2002). *EN, 1990. Eurocode 0: Basis of structural design*. Brussels: European Standards, European Committee for Standardization, European Union.

CEN. (2003). *EN 1991-2. Eurocode 1: Actions on structures - Part 2: Traffic loads on bridges*. Brussels: European Standards, European Committee for Standardization, European Union.

CEN. (2004a). *EN 1992-1-1. Eurocode 2: Design of concrete structures – Part 1-1: General rules and rules for buildings*. Brussels: European Standards, European Committee for Standardization, European Union.

CEN. (2004b). *EN1998. Eurocode 8: Design of structures for earthquake resistance*. Brussels: European Standards, European Committee for Standardization, European Union.

CEN. (2005). *EN 1992-2. Eurocode 2: Design of concrete structures – Part 2: Concrete bridges – Design and detailing rules*. Brussels: European Standards, European Committee for Standardization, European Union.

Gayton, N., Mohamed, A., Sorensen, J.D., Pendola, M., & Lemaire, M. (2004). Calibration methods for reliability-based design codes. *Structural Safety, 26*, 91–121.

HA. (2001). BD 21/01: The assessment of highway bridges and structures. Design manual for roads and bridges. Volume 3. London: Highways Agency.

ISO. (1998). *ISO 2394: General principles on reliability for structures*. Geneva: International Standard Organization.

ISO. (2010). *ISO 13822: Bases for design of structures – Assessment of existing structures*. Geneva: International Standard Organization.

Koteš, P., & Vičan, J. (2013). Recommended Reliability Levels for the Evaluation of Existing Bridges According to Eurocodes. *Structural Engineering International, 23*, 411–417.

Kroon, I. B.. (1994). *Decision theory applied to structural engineering problems* (PhD thesis). Instituttet for bygningsteknik, Aalborg Universitet, Danmark.

MEL (1971). *Fascicule 61, Titre II: Conception, calcul et épreuves des ouvrages d'art – Programme de charges et épreuves des ponts routes* [Book 61, Part II : Design, calculation and load proof tests? Programme for proof load testing]. Paris: Ministère de l'Equipement et de Logement. [in French].

OFROU. (2006). *Evaluation de ponts routiers existants avec un modèle de charge de trafic actualisé*. Estavayes-le-LAc: Office Fédéral des ROUtes.

Sétra. (2011). RelsciTbx: Reliability analysis software (Release 1.0). Sourdun: Sétra.

Sétra. (2012a). Standard methods for structural assessment. Current practice within the Ministry of Transport. Information note, 35. Sourdun: Sétra.

Sétra. (2012b). *CHAMOA – PSIDA: Guidelines for designing reinforced concrete slab bridges*. Sourdun: Sétra.

SIA. (2009). *SIA 269/4: Maintenance of load-bearing structures – Composite steel and concrete structures*. Zurich: Schweizerischer Architekten- und Ingenieurverein.

Sýkora, M., Holický, M., & Marková, J. (2013). Verification of existing reinforced concrete bridges using the semi-probabilistic approach. *Engineering Structures, 56*, 1419–1426.

Performance profiles of metallic bridges subject to coating degradation and atmospheric corrosion

Alexandros N. Kallias, Boulent Imam and Marios K. Chryssanthopoulos

ABSTRACT

This study focuses on deterioration modelling and performance assessment of metallic bridges affected by atmospheric corrosion, considering also the contribution of typical protective systems in the form of multi-layer coatings. The mechanisms leading to coating degradation are reviewed and the main coating types used by infrastructure owners are highlighted. Building on information contained in industry manuals, a simple model for coating degradation is proposed. Atmospheric corrosion models are then presented, with emphasis given to exposure classification, in line with corrosivity classification guidelines and recent research quantifying the influence of corrosion through dose–response functions. Coating degradation and corrosion models are integrated into a modelling framework, aimed at producing performance profiles of elements in metallic railway bridges. Finally, the framework is implemented in a case study in which a range of condition and resistance performance criteria are presented for different elements, such as girders and stiffeners, and their constituent parts, such as webs and flanges. It is shown that the proposed methodology is sufficiently detailed to enable differentiated performance predictions based on key external factors, and has scope for improvement, especially as coating and corrosion models are informed by the collection of field data.

1. Introduction

According to an extensive demographic survey undertaken for the Sustainable Bridges project (Bell, 2007), more than 35% of European railway bridges – whose total number is estimated to be well in excess of 300,000 – are over 100 years old. Although bridges of masonry construction are the single largest group within the overall population, metallic bridges – including those with steel/concrete or encased beam construction – comprise more than a third, with about one in five being centenarian. Steel and wrought iron are the predominant metals, with the latter having been used extensively in the nineteenth century until the era of steel began. In fact, more than 15,000 wrought iron bridges remain in service, mainly on the UK railway network, which had already expanded significantly by the time when steel was mass produced. The same survey, which spanned railway authorities from Italy to Finland and from Poland to Ireland, revealed that the main asset management priorities relate to the improvement of the assessment process, with refined inspection/diagnosis tools and performance prediction also attracting significant attention.

Due to a large proportion of these bridges being on heavily utilised networks, large-scale replacement is practically impossible, notwithstanding the worldwide financial constraints imposed on infrastructure maintenance budgets. Therefore, refined assessment methods and, where necessary, techniques that can improve prediction capability with respect to remaining service lives, thus allowing a life extension even by a modest fraction, are actively being sought. These refined methods could help in improving decision-making related to asset management, and enable an orderly transition to the next generation of railway infrastructure. However, these decisions need to be taken in the context of ageing and exposure to harsh environmental conditions, both of which contribute to wear and deterioration. In the case of metallic bridges, the former manifests itself through fatigue, whereas the latter is principally related to propensity to corrosion; either can have major implications on safety, functionality and appearance. This study focuses on corrosion, which is a commonly observed cause of damage and failure in metallic bridges (Imam & Chryssanthopoulos, 2012; Wardhana & Hadipriono, 2003).

The occurrence and the subsequent rate of progression of the various types of corrosion depend on several factors and their interactions, with the majority associated with exposure conditions and the type of protection system applied, though maintenance practices also play a part. The progression of corrosion damage can lead to inadequate performance and reduced safety, which ultimately can result in structural failures. However, even before structural failure is reached, corrosion deterioration can be assessed as unacceptable on the basis of condition criteria. In simple terms, corrosion can be classified as general or local, though a more precise classification based on the forms of corrosion (e.g. Landolfo, Cascini, & Portioli, 2010) subdivides it into general (uniform), pitting (localised), crevice, erosion, galvanic and fatigue corrosion. It should be noted that over time, a localised region can spread spatially or, conversely, a uniform domain can develop local patterns, emphasising the time-variant complexities of the corrosion process.

Several studies have been carried out dealing with performance assessment of deteriorating steel bridges (e.g. Czarnecki & Nowak, 2008; Kayser & Nowak, 1989; Sharifi & Paik, 2011). However, a systematic exposure classification is hitherto lacking, and corrosion damage is predicted from models based on non-homogeneous databases, typically associated with unqualified levels of uncertainty. Furthermore, in most studies, the time-dependent performance of the protective system, which needs to be broken down before corrosion damage occurs on the metallic surface, has not been considered explicitly.

This paper focuses on deterioration modelling and performance assessment of metallic bridges affected by atmospheric corrosion, considering also the contribution of typical protective systems. Emphasis is given to exposure classification, in line with corrosivity classification guidelines and research aimed at quantifying the influence of corrosion through dose–response functions (DRF). A distinction is made between condition- and resistance-based performance criteria to suit various asset management objectives, which can be addressed through a hierarchical modelling process depending on available information. The proposed methodology is demonstrated using typical metallic bridge elements, for which performance profiles are developed under different short- and long-term exposure scenarios.

2. Deterioration modelling framework

2.1. Exposure classification

Coating deterioration and steel corrosion (following the loss of protection provided by the coating) are time-variant processes, with their rate being determined by the outdoor exposure conditions experienced by the bridge. They can be broadly classified as: (a) immersed, (b) splash zone and (c) atmospheric exposure. Each of these environments is associated with different ranges of corrosivity potentials. This study focuses on atmospheric corrosion, which can itself be subdivided in a number of categories, namely: rural, urban and industrial/coastal exposures (Figure 1). Several studies (e.g. Bs EN I.S.O, 2012a; Feliu, Morcillo, & Feliu, 1993; Gascoyne & Bottomley, 1995; Klinesmith, McCuen, & Albrecht, 2007) have concluded that the main parameters influencing the atmosphere's corrosivity are climatic factors (i.e. relative humidity and temperature) and atmospheric pollutants (i.e. sulphur dioxide SO_2 and chlorides Cl^-). Physical monitoring of these parameters could be used to determine the atmospheric corrosivity classification of a particular location.

The standard Bs EN I.S.O (2012a) provides a framework for the classification of atmospheric corrosivity based on the levels of the main influencing climatic and pollutant variables. Specifically, in BS EN ISO 9223, the spectrum of atmospheric corrosivity is divided into five categories varying from very low (category C1), corresponding to rural environments, up to very high (C5) and extreme corrosivity (CX). In turn, the correlation of these exposure classifiers with metal loss measurements

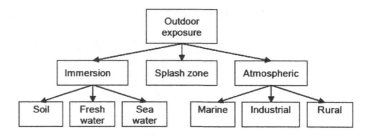

Figure 1. Outdoor exposure environments (Sørensen et al., 2009).

has allowed the development of corrosion models (discussed in Section 2.3). Table 1 briefly describes each corrosivity category together with the expected ranges of corrosion rates within each category.

In considering the deterioration of complex and spatially extended structural systems such as bridges, individual elements (e.g. deck, main girders and cross beams) are likely to experience dissimilar corrosion rates due to differences in the microclimate which develops in their immediate surroundings (Hutchins & McKenzie, 1973). For instance, a Japanese survey has shown that external main girders (EMGs) of steel bridges are more susceptible to corrosion than inner girders (Tamakoshi, Yoshida, Sakai, & Fukunaga, 2006).

2.2. Protective coatings

One of the aims of this study is to examine the performance of coatings applied onto metallic surfaces of bridge components (BS EN ISO, 1998). Several coating types – of varying composition and performance – exist for the protection of structural steelwork, including organic, inorganic, metallic, duplex and hybrid systems (MAINLINE, 2014). In practice, the majority of the coating systems consist of several layers with each layer performing a different function. Organic, metallic and duplex (hybrid) coatings protect the substrate metal through the interface with the anodic and cathodic reactions in the cell and/or by hindering the transport of ions to the substrate surface. Based on their resistance mechanisms, coatings can be classified in the following groups (Greenfield & Scantlebury, 2000; Hare, 2006; de Wit, van der Weijde, & Ferrari, 2011):

- Barrier coatings, which suppress the diffusion of ions through the coating to the substrate surface.
- Inhibitive coatings, which promote the formation of an insoluble passive film on the metal substrate; careful selection of the binder and the inhibitive pigment is required to avoid the rapid exhaust of the inhibitive ions provided by the pigment and reduce blistering.
- Sacrificial coatings, which protect the substrate metal through the mechanism of galvanic corrosion; here, the metal powder in the coating becomes the anode while the substrate metal becomes the cathode.

2.2.1. Coating types

This section highlights the compositions and performance characteristics of the main coating types. Organic coating systems are built-up by several compatible layers, producing a wide range

Table 1. Description of atmospheric corrosivity categories and corresponding corrosion rates (Bs EN I.S.O, 2012a).

Corrosivity category	Description	Corrosion rates[a] (mm/year)
C1	*Very low corrosivity*: Dry or cold zone, atmospheric environment with very low pollution and time of wetness, e.g. certain deserts and Central Arctic/Antarctica	$\leq .0013$
C2	*Low corrosivity*: Temperate zone, atmospheric environment with low pollution ($SO_2 < 5$ µg/m³), e.g. rural areas and small towns. Dry or cold zone, atmospheric environment with short time of wetness, e.g. deserts and subarctic areas	$.0013 < A \leq .025$
C3	*Medium corrosivity*: Temperate zone, atmospheric environment with medium pollution (SO_2: 5 µg/m² to 30 µg/m³) or some effect of chlorides, e.g. urban areas and coastal areas with low deposition of chlorides. Subtropical and tropical zone, atmosphere with low pollution	$.025 < A \leq .050$
C4	*High corrosivity*: Temperate zone, atmospheric environment with high pollution (SO_2: 30 µg/m³ to 90 µg/m³) or substantial effect of chlorides, e.g. polluted urban areas, industrial areas, coastal areas without spray of salt water or and exposure to effect of de-icing salts. Subtropical and tropical zone, atmosphere with medium pollution	$.050 < A \leq .080$
C5	*Very high corrosivity*: Temperate and subtropical zone, atmospheric environment with very high pollution (SO_2: 90 µg/m³ to 250 µg/m³) and/or significant effect of chlorides, e.g. industrial areas, coastal areas and sheltered positions on coastline	$.080 < A \leq .200$
CX	*Extreme corrosivity*: Subtropical and tropical zone (very high time of wetness), atmospheric environment with high SO_2 pollution (higher than 250 µg/m³) including accompanying and production factors and/or strong effect of chlorides, e.g. extreme industrial areas, coastal and offshore areas and occasional contact with salt spray	$.200 < A \leq .700$

[a]Corrosion rates correspond to the first year of exposure.

of properties by varying the composition and thickness of the individual layers. The main constituents of organic coating layers are the binder (usually an epoxy matrix), the pigment and the solvent, with the former determining the physical and chemical properties of the coating. Fillers and additives are commonly used to add specific properties to the final coating, such as impact and abrasion resistance, UV absorption, etc. (Keijman, 1999; de Wit et al., 2011). The first layer, placed in direct contact with the substrate metal, is the primer which can be enriched with corrosion inhibitive pigments, e.g. galvanic pigments incorporating metallic zinc particles (Hare, 2006). The function of the topcoat (or finish) layer is to provide initial resistance against the external environment and determine the aesthetic appearance (e.g. colour).

Metallic coatings provide excellent long-term corrosion protection to metallic structures subjected to a range of harsh exposure conditions (e.g. marine environments). These coatings protect the substrate metal through sacrificial cathodic protection together with barrier action. The metallic coating, typically zinc in hot-dip galvanising or aluminium in thermal-sprayed coatings, acts as the anode of an electrochemical cell (CORUS, 2004). In thermal spraying applications, a heated spray gun (oxygas flame or electric arc) is fed with the metals (powder or wire form) and the melted metal is blown onto the metallic surface using a compressed air jet (CORUS, 2004). Suitable surface preparation facilitates the mechanical bond between the coating and the substrate metal. Guidance on the use of thermal spraying aluminium/zinc coatings can be found in BS EN 22063 (1994).

Duplex (metallic–organic layers) and hybrid (metallic–metallic layers) coating systems provide superior long-term corrosion protection. Typically, duplex coating systems consist of a metallic coating applied directly on the prepared substrate metal followed by subsequent organic coating layers. In this way, the pores of the metallic coating are sealed using the organic coating, adding a barrier against the ingress of corrosive species. In general, the main drawback of duplex and hybrid coating systems is their relatively higher initial cost; often this can be offset by the

anticipated reduced need for both maintenance/repair and minimised service disruption. Hybrid coatings systems, consisting of metal alloys or two dissimilar metallic coating layers, have also been formulated. The results of Kuroda, Kawakita, and Takemoto (2006) from a long-term experimental programme using thermally sprayed and hybrid zinc–aluminium alloy coatings showed that these coatings performed well in marine environments even when unsealed. Similar results were obtained by Salas et al. (2012) using a thermally sprayed two-layer zinc–aluminium hybrid coating system, with the zinc layer applied as a primer.

2.2.2. Coating deterioration

Deterioration of in-service coatings is affected by several factors including coating specification, quality of the application (surface preparation, application and/or curing), the environmental exposure conditions and accidental damage. Clearly, the large number of variables, together with their possible interactions, makes coating deterioration a complex phenomenon. In general, the presence of defects within a coating system accelerates the deterioration process which eventually leads to coating failure.

The most common failure mode is loss of adhesion, for instance due to blistering or cathodic delamination of the coating. Other coating failure criteria include the time needed for the consumption of the active ingredients of inhibitive and sacrificial coatings (i.e. inhibitive pigment and consumption of zinc dust), though the actual time-to-failure is highly uncertain due to the variability in surface preparation, coating application quality, the constantly changing exposure conditions, inherent defects, as well as unexpected factors such as accidental damage. Further information on defects and failure modes of protective coatings can be found elsewhere (e.g. Greenfield & Scantlebury, 2000; Sørensen, Kiil, Dam-Johansen, & Weinell, 2009).

Cathodic delamination is a commonly observed failure mode of organic coatings caused by inherent or induced defects. Corrosion initiates at the defect, forming corrosion products which potentially block the pores. As a result, the area where the coating is damaged becomes the anode while cathodes develop at the edges of the defect. The alkaline environment (pH > 12 due to hydroxyl ions) at the cathodic areas of the substrate is responsible for bond loss at the coating–substrate interface, with the formation of insoluble corrosion products promoting the separation of anodes and cathodes. The rate-controlling parameter in this failure mechanism is the diffusion rate of cations required for charge neutralisation of the cathodically produced hydroxyl ions.

Blisters are typically associated with coatings where no defect is visible. The possible mechanisms involved in this type of failure include expansion due to swelling, gas inclusion or osmotic processes (Sørensen et al., 2009). Among these mechanisms, osmotic processes have been shown to be the most significant in promoting blistering. Greenfield and Scantlebury (2000) have studied the development of blisters, which have been classified as osmotic, anodic and cathodic.

2.2.3. Coating selection

In the UK, corrosion protection of railway bridges using coatings and sealants is covered by guidance documents produced by the infrastructure owner. Current recommendations distinguish between new and existing steelwork, with a number of coating systems recommended for each category (see NR, 2009a, 2009b, 2009c). Commonly used coatings for new construction include duplex systems in which the first of several layers is a metallic coating, for instance thermally sprayed zinc or aluminium. On the other hand, coating systems for maintenance of existing structures are broadly subdivided into those recommended for patch repairs (bitumen based) and those for complete reapplication. These protective systems typically consist of several layers applied sequentially on the treated steel surface.

As mentioned above, the performance of a coating system is influenced by several factors, including the composition and exposure conditions, as well as workmanship and quality control during application. For example, Table 2 summarises the characteristics and expected service life of two coatings used in railway maintenance. Guiding values for the expected service life for other coating systems currently used are given in the aforementioned manuals (NR, 2009a, 2009b, 2009c). It is also recommended that the selection of a suitable protective system be related to the anticipated exposure conditions (e.g. classification of exposure conditions based on BS EN ISO 12944-2).

2.2.4. Coating deterioration modelling

In general, the development of models to predict coating performance requires direct or indirect consideration of the following parameters:

- The coating's chemical and physical characteristics, including its principal resistance mechanism (e.g. barrier, inhibitive and galvanic action).
- The anticipated exposure conditions, including levels of individual climatic and atmospheric variables for a particular location.
- Other aspects such as the quality of surface preparation, operator's skills and experience and coating curing characteristics.

Thus, modelling of coating performance can be attempted at different levels of complexity and accuracy, as outlined below:

- In Level 1 models, either a single value or a range is available for the expected service life of a coating system under a particular environmental exposure; no quantitative information is available on the influence of individual climatic or atmospheric factors on coating performance. Such models may include, based on inspection data, the primary statistical properties of the coating's service life (i.e. mean and standard deviation), which could reflect not only the influence of different exposure conditions (i.e. rural and urban) but also the variability in workmanship (e.g. poor and good).
- Level 2 models are based on mathematical relationships between coating performance and a set of influencing variables (e.g. temperature and humidity) known as DRF. Where the statistical properties of the influencing factors are also available, the corresponding coating performance statistics may be inferred using expectation properties or Monte Carlo simulation.
- Level 3 models are based on theoretically underpinned analytical and numerical simulations. As opposed to the DRFs, the underlying formulations describe the mechanisms involved in the deterioration processes (Nguyen, Hubbard, & Pommersheim, 1996; Pommersheim, Nguyen, Zhang, & Hubbard, 1994). These models can improve fundamental understanding through the interpretation of experimental data. Currently, their use is limited to carefully executed laboratory studies.

In the following, a simple (Level 1) model for time-dependent coating performance is developed, using information available in Network Rail standards where a range of expected life (T_L) values are provided for a number of coatings used in the UK (see NR, 2009a, 2009b, 2009c). Experimental results (Itoh & Kim, 2006; Kim & Itoh, 2007) suggest that the coating itself will gradually deteriorate with time and will eventually become

Table 2. Description and expected service life of protective coatings (NR, 2009a, 2009b).

Coating system name	Description [for more details see (NR, 2009a, 2009b)]	Expected life, T_L (years)
M27.4	*Protective system using bitumen.* First layer: surface tolerant epoxy primer (min dft 100 μm. Intermediate coat: gelled bituminous solution – aluminium tinted (min dft 200 μm). Topcoat: gelled bituminous solution – black finish with min dft 200 μm	5–7
M21	*Coating using epoxy glass flake.* First layer: epoxy blast primer with min dft 25 μm. Intermediate layer: epoxy glass flake intermediate coat with min dft 40 μm and for layer C select among the following (min dft 50 μm): anti-graffiti paint or polyurethane-coloured finish or acrylic urethane topcoat or polysiloxane topcoat	18–22

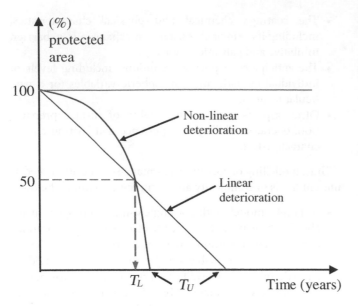

Figure 2. Level 1 deterioration model for coating system performance.

completely ineffective. This assumes that the application of the coating is carried out by competent coating contractors, and, hence, the possibility of rapid deterioration and/or premature peeling is minimised. Furthermore, the results of Itoh and Kim indicate that coating deterioration tends to be non-linear over time, as schematically depicted in Figure 2. This is further supported by a study on the quality of steel bridge coatings (Chang, Georgy, & AbdelRazig, 2000) in which non-linear deterioration curves similar to those shown in Figure 2 are proposed based on the relationship between warranty period and expected service life. Building on these idealisations, a polynomial relationship is proposed, assuming that service life (T_L) estimates quoted in industry manuals correspond to ca. 50% of a coated surface being unprotected at time T_L:

$$\frac{A_{pr}(t)}{A_{pr0}} = 1 - \left(\frac{0.6t^2}{T_L^2} - \frac{0.1t}{T_L} \right) \qquad (1)$$

where $A_{pr}(t)$ and A_{pr0} are the residual and initial protected areas, respectively, and t is the time in years. This equation may be used to predict the time T_U at which the coating is lost from the entire surface area. Note that $A_{pr}(t) \leq A_{pr0}$ for all t and $A_{pr}(t) = 0$ for $t > T_U$.

In contrast to a linear relationship (also shown in Figure 2), for which $T_U = 2T_L$, the proposed polynomial leads to $T_U = 1.38$ T_L. Given the observed trends in experimental and field studies, this is a first attempt at a coating performance model, which can be used to estimate deterioration over any given surface (e.g. a web and a flange of a plate girder), for situations where the deterioration is likely to become progressively more extended. It should be noted that the coefficients in Equation (1) could be adjusted to reflect alternative values regarding the percentage coating loss taken as the criterion estimating expected service life. In general, such a model should consider coating performance not only at a particular location but also its spatial characteristics. To enable this, information on whether uniform or localised corrosion is likely to be developed on a particular member should also be utilised.

2.3. Atmospheric corrosion models

At a specimen scale, the progression of atmospheric corrosion in any given environment can be modelled in terms of thickness loss over an exposed surface area. Over the years, models of varying complexity and accuracy have been developed, as summarised in Table 3. The proposed classification accords with that adopted for coating deterioration in Section 2.2.4.

Among Level 1 models, the best known is the power model (Feliu et al., 1993;) given by:

$$C(t) = At^B \qquad (2)$$

where, $C(t)$ is the uniform (measured as average over relatively small specimen surface areas) thickness loss (mm) after an exposure period of t years and coefficients A (mm/year) and B (unitless) are empirical constants, obtained using regression analysis on physical test results grouped according to different atmospheric exposure conditions. This implies that the dependency on the exposure conditions and material type (e.g. type of steel) is captured implicitly using suitable values for coefficients A and B, where A represents the corrosion loss at the end of the first year and B controls the rate of loss in subsequent years. Recommended values for A and B are generally deemed to exhibit high uncertainty, partly as a result of databases in which exposure conditions were poorly defined/grouped. It has also been suggested that Equation (2) is suitable for periods up to 20 years; for $t > 20$ years, thickness loss may be calculated using the linear relationship presented below (Bs EN I.S.O, 2012b):

$$C(t > 20) = A\left[20^B + B\left(20^{B-1} \right)(t - 20) \right] \qquad (3)$$

Figure 3 depicts the thickness loss over time, for different corrosivity categories, predicted using Equations (2) and (3), with coefficients A and B chosen as either mid-range or upper bound values. The results indicate that for corrosivity categories C1 to C4, the predicted corrosion losses are less than about .5 mm over a 30-year period, if no protective system is applied. On the other hand, much higher losses (more than twofold) are predicted for corrosivity category C5. The range in the values of the empirical coefficients A and B, indicative of the inherent uncertainty, can also be seen to affect the predicted losses.

Level 2 models offer the opportunity to relate directly the environmental and atmospheric pollutant variables to the rate of corrosion, instead of pooling the influence of all factors on empirical model constants. With the availability of data pertaining to environmental variables and/or atmospheric pollutant concentrations increasing (partly as a result of raised concerns over air quality in many urban/industrial areas), these more advanced models could, in future, offer refined corrosion predictions, potentially also allowing changes in input variables over time to be accommodated.

An example of a relationship, which can be used to estimate coefficient A in Equation (2) as a function of several environmental and atmospheric pollutant variables, leading to a Level 2 corrosion model, is given in Bs EN I.S.O (2012a):

$$A = 1.77SO^{0.52} \exp(0.02RH$$
$$+ f_{St}) + 0.102Cl^{0.62} \exp(0.033RH + 0.04T) \qquad (4)$$

Table 3. Corrosion model classification.

Level	Description	Comments
1	Empirical models. Effect of all influencing factors taken into account through model constants	Model coefficients are associated with very high uncertainty. Statistical properties may not be reliable due to inhomogeneous samples. Questionable model transferability
2	Empirical models which relate the rate of corrosion to specific exposure variables, also known as dose response functions (DRF)	Reliance on spatial data for atmospheric pollutants and climatic parameters. Potentially suitable for probabilistic analysis, though uncertainty modelling largely untested
3	Theoretical models involving simulation techniques to predict airflow patterns and pollutant mass transfer on exposed surfaces	Heavy reliance on modelling assumptions. Complex uncertainty modelling, currently lack of input data at desired granularity level

where SO is the average annual deposition of SO_2 (mg/m² d), RH is the average annual relative humidity (%), Cl is the average annual deposition of Cl^- (mg/m² d), T is the annual (average) temperature (°C) and f_{St} is a function of T.

An alternative Level 2 model has been presented by Klinesmith et al. (2007):

$$C(t) = A' \left(\frac{TOW}{C_1} \right)^D \left(1 + \frac{SO_2}{E} \right)^F \left(1 + \frac{Cl}{G} \right)^H e^{J(T+T_0)} t^B = A_{KL} t^B \tag{5}$$

where TOW is the time of wetness (h/year), SO_2 is the sulphur dioxide concentration (µg/m³), Cl is the chloride deposition rate (mg/m² d), $C_1 = 3800$ h/year (mean value), E = mean of measured values of SO_2, T is the air temperature (°C) and $T_0 = 20$ °C. For flat carbon steel specimens (Klinesmith et al., 2007): $A = 13.4$, $B = .98$, $D = .46$, $F = .62$, $H = .34$ and $J = .016$. Comparing Equations (4) and (5), it can be observed that both models retain the exponential term (t^B) but differ in the modelling approach for the corrosion loss in the first year (represented by the remaining part of the function).

Figure 4(a)–(d) presents a sensitivity analysis on the effect of different climatic and atmospheric variables on the long-term predictions of corrosion thickness losses calculated using Equation (5). For the variables involved, typical values are used, as suggested in Bs EN I.S.O (2012a). As can be seen, airborne salinity and SO_2 concentration have the greatest impact on the long-term rate of corrosion.

As mentioned earlier, Level 2 models could also capture the potential effects of changes to exposure conditions over longer time horizons. These may be associated with initiatives to reduce atmospheric pollution in specific regions or related to effects brought about by climate change. For this purpose, Equation (3), which is recommended for estimating losses beyond 20 years, is recast in the following form:

$$C(t < 20) = A \cdot 20^B + A \cdot B \left(20^{B-1} \right)(t - 20)$$
$$= C(t = 20) + \frac{dC(t)}{dt}(t - 20) \tag{6}$$

where the first term represents the thickness loss due to corrosion over 20 years and the second term is the thickness loss from 20 years onwards. As can be seen, the first term (accounting for corrosion up to $t = 20$) is of exponential form, whereas the second term (accounting for corrosion when $t > 20$) is linear with a slope equal to $dC(t)/dt = A \cdot B(20^{B-1})$. In the case where there is a change in exposure conditions at a time $t > 20$ years, the slope of the linear portion $dC(t)/dt$ can be modified accordingly, by defining a new A_i coefficient to account for the new exposure

conditions. Thus, for n changes of exposure conditions beyond $t = 20$ years, the following expression may be used:

$$C(t > 20) = C(t = 20) + \sum_{i=1}^{n} \frac{dC_i(t)}{dt} T_i \tag{7}$$

where T_i represents the duration over which each A_i is valid after the initial 20-year period.

Figure 5 illustrates the corrosion loss predictions for two scenarios assuming atmospheric pollution levels in a specific area are reduced, together with a no-change case. The first scenario assumes a sharp reduction in the SO_2 concentration at year 20 from 250 to 125 µg/m³, whereas in the second scenario, this reduction takes place gradually between year 20 and 40. Both emission reduction scenarios can be seen to result in lower thickness losses, ca. 1 to 1.5 mm, compared to the no-change case. Note that the lines for reduced emission scenarios become parallel after year 40 since the pollutants converge to the same level.

3. Performance assessment

The consequences of deterioration on metallic bridge elements can vary from aesthetic and non-structural issues to progressive weakening and catastrophic failures (Prucz & Kulicki, 1998). In the context of maintenance regimes, the former are addressed through condition surveys, whereas the latter necessitate structural assessments. In this section, the coating and corrosion models presented above are, first, combined so as to provide predictions of coating and material loss under different exposure conditions (condition-based maintenance) and, secondly, integrated within strength formulations relevant to different limit states, such as tension, compression, bending and shear (capacity-based maintenance).

In general, environmental exposure leads initially to a breakdown and loss of coating over a growing number and extent of surface spots, and, while this process is continuing, corrosion also begins to take place in the least protected and more vulnerable spots. As previously mentioned, the progression of deterioration will be influenced significantly by the type of coating and substrate material, the presence of pollutants, bacteria and microclimate effects regarding temperature, humidity, wind direction, exposure to sunlight, etc. Hence, prediction of deterioration rates will be characterised by uncertainty and complexity at different scales. In this respect, the proposed modelling approach cannot predict accurately the deterioration of a single structure unless it is combined with inspection outcomes, which is beyond the scope of the present study. Nevertheless, it can provide an insight into median trends that may be observed over time in cohorts of similar structures.

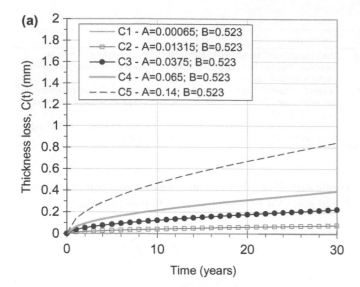

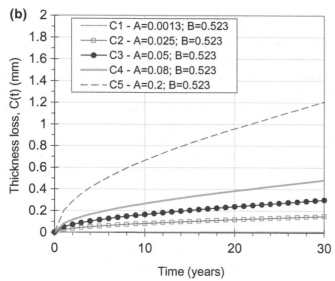

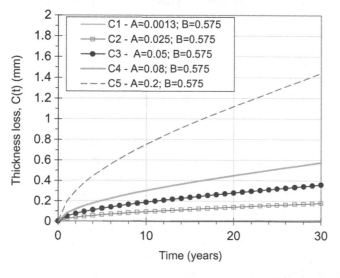

Figure 3. Thickness loss over time predicted using Equations (2) and (3) and Table 1: (a) mid-range values for A; $B = .523$, (b) upper limits for A; $B = .523$ and (c) upper limits for A; $B = .575$.

Figure 6 shows schematically the proposed modelling approach for performance assessment. The coating performance is captured via Equation (1), with the service life, T_L, estimated

based on information on particular coatings, e.g. as found in maintenance manuals. In estimating T_L, the factors shown in Figure 6(a), (specification, application quality and exposure) should, where possible, be taken into account. Material loss due to corrosion will then start at different points in time on the exposed surface, as shown in Figure 6(b); for this part, either Level 1 (Equation (2)) or Level 2 (Equations (4) or (5)) corrosion models may be used, depending on available information on exposure conditions.

The loss of thickness due to corrosion over any particular surface of a single member (e.g. web and flange plates in Figure 6(b)) will have an impact on the available cross-sectional area, which, in turn, influences other section properties relevant to different limit states ((bending, compression, etc.), e.g. the second moment of area or the radius of gyration. Thinning of the web or flange plate can also affect adversely local buckling strength, which is typically a function of the member slenderness, e.g. the flange half-width over thickness ratio. Moreover, if corrosion is allowed to progress considerably over an entire structure, overall load distribution may be affected due to changes in members' axial and bending stiffness. Following the approach suggested by Prucz and Kulicki (1998), a number of residual resistance factors are proposed in the following, covering the principal limit states encountered in structural assessments of girder and truss bridges. Thus, for a bending limit state, this factor is given by:

$$R_B(t) = \left[\frac{S(t)}{S_0} \right] \quad (8)$$

where, $R_B(t)$ is the bending resistance ratio and S_0 and $S(t)$ are the initial (uncorroded) and the time-varying (corroded) values of section modulus, respectively. Depending on the class of the section (i.e. whether the elastic or plastic bending capacity can be reached), either the elastic or plastic section modulus would have to be considered (Bs EN 1993-1-1, 2005). Clearly, for the calculation of the section modulus at time t, the distribution of the induced corrosion damage across the section needs to be idealised; here, unless specific scenarios based on microclimate effects are considered, the baseline assumption is that the loss of material is spread uniformly across the width of the flanges and along the web depth. However, it is possible to adapt this equation to reflect the situation (sometimes observed in real cases) where the bottom flange suffers more than the top flange due to local water entrapment, bird fouling, etc.

If shear capacity is of concern, the corresponding ratio is given by:

$$R_S(t) = \left[\frac{t_w(t)}{t_{w0}} \right] \quad (9)$$

where $R_S(t)$ is the shear resistance ratio and t_{w0} and $t_w(t)$ are the thicknesses of the uncorroded and corroded web elements, respectively.

Instabilities may be an issue either in the form of member (Euler) buckling or in the form of local buckling, e.g. in flange outstands and slender webs. For member buckling, two ratios are relevant, depending on whether the member falls in a stocky or slender category. For the former, the cross-sectional area becomes the controlling geometric property since material failure precedes buckling, thus the buckling ratio becomes:

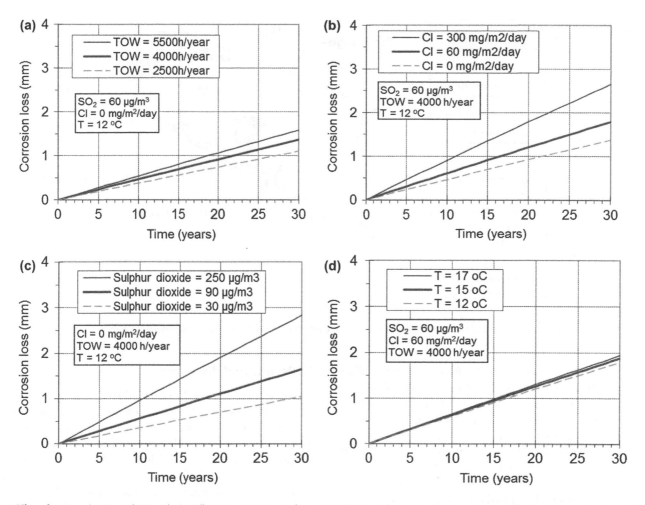

Figure 4. Effect of varying climatic and atmospheric pollutant parameters on the corrosion losses with time, predicted using Equation (5): (a) time-of-wetness (TOW), (b) Cl deposition rate, (c) SO$_2$ concentration and (d) temperature.

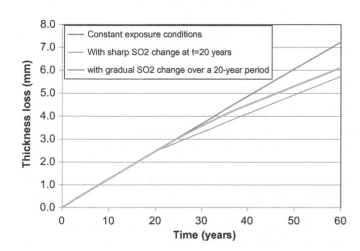

Figure 5. Thickness loss over time under changing exposure conditions.

$$R_A(t) = \left[\frac{A(t)}{A_0} \right] \quad (10)$$

with A_0 and $A(t)$ being the uncorroded and corroded areas. For slender members in compression governed by elastic buckling, the ratio changes to:

$$R_{EB}(t) = \left[\frac{I(t)}{I_0} \right] \quad (11)$$

where, I_0 and $I(t)$ are the second moments of area about the relevant member axis (major/minor).

As for local buckling, the available resistance, as determined via plate buckling formulae (Timoshenko, 1964), is proportional to the square of the thickness, and therefore an appropriate local buckling ratio is given by:

$$R_{LB}(t) = \left[\frac{t_{fl}(t)}{t_{fl,0}} \right]^2 \quad (12)$$

where, $t_{fl,0}$ and $t_{fl}(t)$ are the thicknesses of the compression flange (or any other element).

Finally, it is worth noting that for members under tension, the capacity is once more governed by the cross-sectional area and the relevant ratio is, thus, given by Equation (10).

4. Case study

Following the performance modelling framework presented in the preceding section, performance profiles are presented for members of a short-span (half-through) railway bridge with a

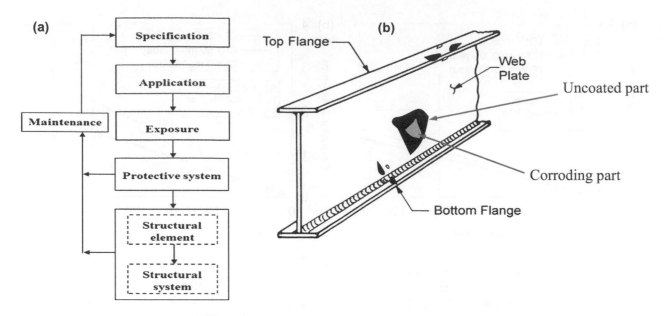

Figure 6. Performance modelling framework: (a) factors under consideration (b) progression of deterioration on coated metallic surfaces.

span of 9.6 m, as shown in Figure 7, located in a heavily polluted industrial site (C5 corrosivity classification as per Table 2). The examined bridge, which is typical on the UK railway network, consists of different member types, including external/internal main girders, stringers and cross beams. The EMGs have 13-mm thick top and bottom flanges and 10-mm thick webs, with the overall height of the section being 1220 mm. The yield strength of the material is taken as f_y = 300 MPa. It is further assumed that a protective coating is applied initially but no coating reapplication takes place during the examined 30-year maintenance planning window. However, the framework can also be used for an uncoated structure to focus on the effect of the coating on the structural performance, as well as to include the effect of coating reapplication within a given maintenance period.

As often observed in inspections, the actual position of the bridge members (i.e. internal/external) can influence the aggressiveness of the microclimate to which their metal surfaces are exposed. In one such scenario, the exposure conditions faced by internal members are less harsh compared to external members: thus, whereas the global exposure classification matches the microclimate of external members, internal members (e.g. stringers) are exposed to microclimates of lower aggressiveness. As mentioned previously, this assumption can be further refined, should particular factors prevail that may distinguish sub-elements of a girder, either in a vertical direction (e.g. top vs. bottom flange) or horizontally (e.g. end vs. middle sections).

Table 4 lists the scenarios for which performance profiles are developed. The service life of the coatings (T_L) is taken as the mid-range of the values given in Table 2. In view of the bridge type, bending, shear and local buckling limit states are considered for different members. In particular, overall bending and shear are examined for the internal and external girders, whereas local buckling is assessed for the compression flange of the same members. For completeness, other members (e.g. cross-girders and longitudinal stiffeners) could also be included, bearing in mind the particular limit states that may govern their performance.

5. Results and discussion

In this section, the results obtained from the case study which illustrate the methodology are presented and discussed in relation to the observed performance for different coating types, exposure conditions and location of different elements within a typical metallic railway bridge. Combining condition-based indicators (e.g. % area of coating breakdown) with the evolution of structural resistance can be a useful tool within the context of asset management. Specifically, understanding the effect of different exposure conditions on the expected evolution of condition and/or resistance over time would allow the optimisation of resource allocation and justify the prioritisation of future examinations and assessments while maintaining the overall risk associated with a bridge portfolio at tolerable levels.

5.1. Condition assessment

5.1.1. Coating performance

Figure 8 shows the evolution of coating performance predicted using Equation (1) for two coating systems subjected to two different atmospheric exposure conditions. The results indicate that for all cases examined, an initial period exists during which the coating remains intact, irrespective of the exposure conditions. However, the higher performance coating M21 is associated with a much longer period during which it remains fully effective; in fact, almost double the period estimated for the M27.4 coating. Moreover, the effect of different exposure conditions becomes quite significant beyond this initial post-application period.

For example, for coating M27.4, it can be seen that when 50% of the substrate metal area is unprotected under C3 exposure conditions, only approximately 25% of the area remains protected when considering the C5 environment. Such results can be used to facilitate decision-making within the context of long-term maintenance and whole life cost assessment (MAINLINE, 2013a, 2013b, 2013c). To this end, the collection of field data from regular inspections would allow the improvement of

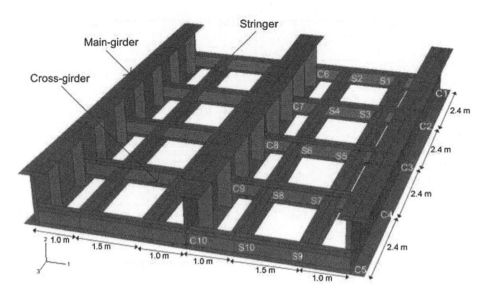

Figure 7. Schematic view of the short-span railway bridge.

Table 4. Scenarios considered in the case study.

Element type[a]	Exposure classification[b]		Coating	
	Bridge level	Element level	Coating type	Service life[c], T_L (years)
IMG	C5	C3	No Coating	–
IMG	C5	C3	M27.4	6
IMG	C5	C3	M21	20
EMG	C5	C5	No Coating	–
EMG	C5	C5	M27.4	5
EMG	C5	C5	M21	18
ST	C5	C3	No Coating	–
ST	C5	C3	M27.4	6
ST	C5	C3	M21	20

[a]IMG = Internal main girder, EMG = External main girder and ST = Stringer beam.
[b]Environmental classification in line with BS EN ISO 9223.
[c]Expected coating life based on NR/GN/CIV/002. Coating specifications are given in Table 2.

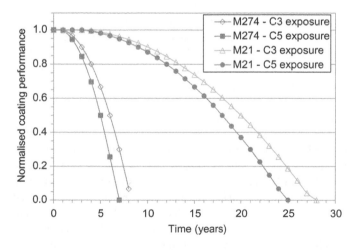

Figure 8. Performance profiles of two coating systems under different exposure conditions.

coating deterioration models, whereas simultaneous recording of the geographical locations would facilitate estimation of statistical properties. At present, although inspection data are being routinely collected, there is no attempt to correlate these

with atmospheric conditions or positioning data, which hinder model development.

5.1.2. Thickness losses

Figure 9(a) and (b) shows the predicted flange thickness and web losses over time for external girders (EMG) subjected to C5 exposure conditions. In all cases, an initial period exists during which coatings are fully effective and no thickness loss is recorded. Thereafter, gradual breakdown of the coating leads to thickness loss over time with the rate of thickness loss being a function of several environmental and pollution-related variables (e.g. temperature, humidity, SO_2 and Cl). As previously discussed, their effect is collectively expressed through a particular set of A and B coefficients.

In Figure 9(a) and (b), results are also shown for the case where no coating is applied to the exposed surfaces; in this case, the residual thickness of the flanges and web (or alternatively loss of thickness) follows a non-linear curve which gradually becomes linear for $t > 20$ years. Figure 10(a) and (b) shows that similar results are obtained for the internal main girder (IMG) element; however, the predicted reductions of thickness are of much smaller magnitude due to the less aggressive atmospheric environment (C3) considered relevant for these elements, which

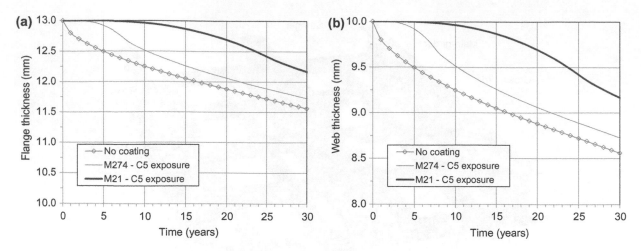

Figure 9. Residual thickness profiles for EMG elements under C5 exposure conditions: (a) flanges thickness and (b) web thickness.

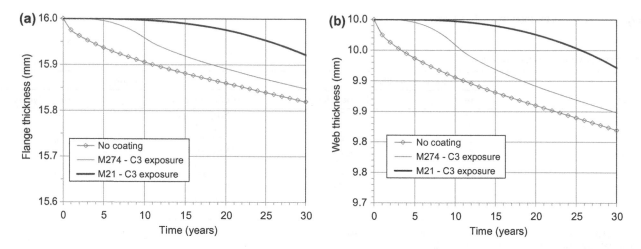

Figure 10. Residual thickness profiles for IMG elements under C3 exposure conditions: (a) flanges thickness and (b) web thickness.

has a dual benefit: (a) longer coating protection and (b) smaller corrosion rates. The assumption differentiating member exposure in a particular structure was based on surveys indicating the higher susceptibility to atmospheric corrosion of external bridge members, in comparison to their internal counterparts (Tamakoshi et al., 2006). It is important to note that microclimate effects (e.g. water ingress and inaccessible/hidden parts) can reverse this trend and reveal higher corrosion in internal parts.

5.2. Resistance assessment

Performance profiles of deteriorating structures can be evaluated with respect to different serviceability and ultimate limit state criteria, for instance increasing deflections, fatigue, bending resistance, shear resistance, local and global buckling resistances. In this paper, results are presented for the bending, shear and local buckling resistances of the different element types (e.g. EMG and IMG) considering their relative location on the bridge as well as different types of protective coatings.

Figure 11 displays a comparison of bending performance profiles obtained using Equation (8) for EMG and IMG elements considering two different coating types, with the no-coating case also shown. The results indicate that, although the bridge is in a

high corrosivity environment, different exposure conditions at element level (due to their relative position on the bridge) cause the bending resistance of EMG elements to deteriorate at a much faster pace relative to the IMG element. The shear performance profiles (through Equation (9)) in Figure 12 follow a similar trend to the bending results; although in this case, the reduction in shear resistance is slightly more severe.

For example, the results in Figures 11 and 12 indicate that at the end of the 30-year maintenance window, the shear and bending resistances of EMG with coating M274 have reduced by more than 12 and 10%, respectively. Figure 13 shows a comparison of the results for the local buckling performance profiles calculated using Equation (12). These results indicate that the reduction in buckling resistance over time occurs at a much higher rate relatively to the bending and shear resistances. Overall, the results indicate that the deterioration of all performance metrics examined is moderate for an initial 10-year period, irrespective of the exposure conditions and coating type, with the reductions being practically within 5%.

However, after this initial period, significant differences in performance are predicted for the examined scenarios. For all performance metrics, the results highlight the significance of the relative location of an element within a bridge, as well as

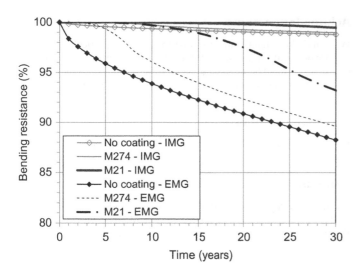

Figure 11. Bending resistance profiles for EMG and IMG elements under, respectively, C5 and C3 exposure conditions.

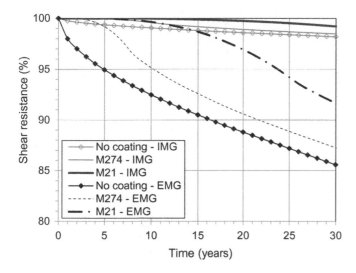

Figure 12. Shear resistance profiles for EMG and IMG elements under, respectively, C5 and C3 exposure conditions.

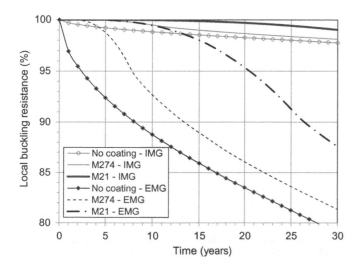

Figure 13. Local buckling resistance profiles for EMG and IMG elements under, respectively, C5 and C3 exposure conditions.

the impact of the protective system on the progression of deterioration. Similar trends were observed in the results for the stringer (ST) elements, which – as for the IMG elements – are classified as internal elements and are, thus, exposed to lower corrosivity (C3).

5.3. General remarks

As discussed earlier, decision-making in asset management can be based on condition and/or resistance criteria, which often complement each other. To this end, it is not unlikely to encounter cases where the gradual loss of condition (e.g. due to the breakdown of the protective coating system) is not associated with loss of resistance, at least during the early stages of the deterioration process. As such, considering both types of performance metrics within a common framework can inform risk-based examination and assessment regimes (e.g. Barone, Frangopol, & Soliman, 2014; Zonta, Zandonini, & Bortot, 2007). Furthermore, such a framework is an integral part of any life-cycle cost methodology (MAINLINE, 2013a, 2013b, 2013c).

Differences between the progression of deterioration in relation to condition and structural performance are often directly related to the exposure conditions, as well as the coating type and its quality. Often, the effect of microclimate associated with specific detailing or other characteristics of a structure can be a significant factor in the evolution of both the condition and structural performance of an element (van de Lindt & Ahlborn, 2005). In this paper, effects arising from the different atmospheric exposure conditions are only considered; the occurrence and evolution of defects associated with local microclimates caused by unsuitable detailing (e.g. water traps, elements buried in ballast and timber–metal interfaces) on a structure have not been considered (Ahn, Kim, Kainuma, & Lee, 2013; Khurram, Sasaki, Kihira, Katsuchi, & Yamada, 2014; van de Lindt & Ahlborn, 2005).

In practice, it is possible to observe significant localised defects associated with such detailing on structures located in relatively benign exposure conditions and which are otherwise in reasonable condition. Notwithstanding, the results of the case study demonstrate that the proposed framework is sufficiently detailed to differentiate performance predictions based on key external factors and has room for improvement, especially as coating and corrosion models are informed by the collection of field data. Finally, it is relatively straightforward to introduce uncertainty modelling on a number of key variables and to estimate the ensuing statistical properties of the performance profiles, in terms of median or appropriate fractile values.

6. Conclusions

A methodology has been presented for the development of performance profiles for deteriorating elements in metallic bridges. The gradual breakdown of the protective coating and the effects of atmospheric corrosion have both been modelled, taking into account their dependency on environmental exposure conditions and a variety of other factors. An approach for long-term changes in pollution or climatic variables has also been proposed. The methodology has been demonstrated through generation of performance profiles for a short-span metallic railway bridge, assumed to be located in a harsh environment.

It is shown that it can account for a wide range of exposure conditions, and that it can be adapted so as to cater for microclimate effects both at inter- and intra-element levels. Depending on policy, budget and maintenance constraints, ageing metallic bridges may have to be managed either using condition or resistance criteria. The developed methodology has the flexibility to enable both to be examined, using relatively simple models which have been based on industry manuals and international guidance documents.

Nomenclature

A	is an empirical constant depending on exposure conditions (mm/year)
A_0	is the uncorroded area
$A(t)$	is the corroded area
$A_{pr}(t)$	is the residual protected area of coated surface
A_{pr0}	is the initial protected area of coated surface
B	is an empirical constant depending on exposure conditions
$C(t)$	is the uniform thickness loss (mm)
Cl	is the (annual average) chloride deposition rate (mg/m^2 d)
f_{St}	is a function of T
I_0	is the uncorroded second moment of area about the relevant member axis (major/minor)
$I(t)$	is the corroded second moment of area about the relevant member axis (major/minor)
$R_A(t)$	is the buckling ratio
$R_B(t)$	is the bending resistance ratio
$R_{EB}(t)$	is the elastic buckling ratio
RH	is the average annual relative humidity (%)
$R_{LB}(t)$	is the local buckling ratio
$R_S(t)$	is the shear resistance ratio
S_0	is the initial (uncorroded) value of section modulus
SO	is the average annual deposition of SO_2 (mg/m^2 d)
SO_2	is the sulphur dioxide concentration ($\mu g/m^3$)
$S(t)$	is the time-varying (corroded) value of section modulus
t	is the time (years)
T	is the (annual average) air temperature (°C)
$t_{fl,0}$	is the uncorroded thicknesses of the compression flange (or any other element)
$t_{fl}(t)$	is the corroded thicknesses of the compression flange (or any other element)
T_L	is the service life
TOW	is the time of wetness (h/year)
T_U	is the time at which the coating is lost from the entire surface area
t_{w0}	is the thickness of the uncorroded web element
$t_w(t)$	is the thickness of the corroded web element

Acknowledgements

We would like to thank our project partners and, in particular, Brian Bell and David Castlo from Network Rail (UK) for fruitful discussions. The opinions expressed herein are those of the authors only and should not be taken as representative of any of the organisations involved.

Disclosure statement

No potential conflict of interest was reported by the authors.

Funding

Part of the work presented in this paper has been undertaken in the course of the FP7 European Union funded project MAINLINE (Maintenance, Renewal and Improvement of Rail Transport Infrastructure to reduce Economic and Environmental Impacts).

References

Ahn, J. H., Kim, I. T., Kainuma, S., & Lee, M. J. (2013). Residual shear strength of steel plate girder due to web local corrosion. *Journal of Constructional Steel Research, 89*, 198–212.

Barone, G., Frangopol, D., & Soliman, M. (2014). Optimization of life-cycle maintenance of deteriorating bridges with respect to expected annual system failure rate and expected cumulative cost. *Journal of Structural Engineering, 140*, 04013043.

Bell, B. (2007). How the project priorities were established. In *Sustainable Bridges, Assessment for Future traffic Demands and Longer Lives*, edited by J. Bień, L. Elfgren, & J. Ol-ofsson, Dolnośląskie Wydawnictwo Edukacyjne.

Bs EN 1993-1-1. (2005). *Eurocode 3: Design of steel structures – Part 1-1: General rules and rules for buildings*. London: British Standards Institute.

BS EN 22063. (1994). *Metallic and other inorganic coatings. Thermal spraying. Zinc, aluminium and their alloys*. London: British Standards Institute.

BS EN ISO. (1998). *BS EN ISO 12944-2. Paints and varnishes – Corrosion protection of steel structures by protective paint systems – Part 2: Classification of environments*. London: British Standards Institute.

Bs EN I.S.O. (2012a). *BS EN ISO 9223:2012 – Corrosion of metals and alloys – Corrosivity of atmospheres – Guiding values for the corrosivity categories*. London: British Standards Institute.

Bs EN I.S.O. (2012b). *BS EN ISO 9224:2012 – Corrosion of metals and alloys – Corrosivity of atmospheres – Classification, determination and estimation*. London: British Standards Institute.

Chang, L. M., Georgy, M. E., & AbdelRazig, Y. (2000). Warranting quality of steel bridge coating. *Journal of Construction Engineering and Management, 126*, 374–380.

CORUS. (2004). *The prevention of corrosion on structural steelwork*. London: Corus Construction & Industrial.

Czarnecki, A. A., & Nowak, A. S. (2008). Time-variant reliability profiles for steel girder bridges. *Structural Safety, 30*, 49–64.

de Wit, J. H. W., van der Weijde, D. H., & Ferrari, G. (2011). Organic coatings. In Marcus P. (Ed.), *Corrosion Mechanisms in Theory and Practice*. (3rd ed., pp. 863–905). London: CRC Press.

Feliu, S., Morcillo, M., & Feliu, S., Jr. (1993). The prediction of atmospheric corrosion from meteorological and pollution parameters – I. *Corrosion Science, 34*, 403–414.

Gascoyne, A., & Bottomley D. (1995). Atmospheric corrosion rates of railway bridge structures. *Report No. LR-MSU-084*. Issued by Scientifics for the British Rail Research British Railways Board, UK.

Greenfield, D., & Scantlebury, D. (2000). The protective action of organic coatings on steel: A review. *The Journal of Corrosion Science and Engineering, 3*, Paper 5.

Hare, C. H. (2006). Corrosion and its control by coatings. *Coatings Technology Handbook, Taylor & Francis Group, 102*, 1–9.

Hutchins, J. S., & McKenzie, M. (1973). *Characterisation of bridge locations by corrosion and environmental measurements – First year results* (Report No. LR550). Crowthorne: Transport and Road Research Laboratory.

Imam, B. M., & Chryssanthopoulos, M. K. (2012). Causes and consequences of metallic bridge failures. *Structural Engineering International, 22*, 93–98.

Itoh, Y., & Kim, I. T. (2006). Accelerated cyclic corrosion testing of structural steels and its application to assess steel bridge coatings. *Anti-Corrosion Methods and Materials, 53*, 374–381.

Kayser, J. R., & Nowak, A. J. (1989). Capacity loss due to corrosion in steel-girder bridges. *Journal of Structural Engineering, 115*, 1525–1537.

Keijman, J. M. (1999). Inorganic and organic coatings – The difference. PCE' 99 Conference: Achieving quality in coatings work: the 21st Century challenge. Brighton, UK.

Khurram, N., Sasaki, E., Kihira, H., Katsuchi, H., & Yamada, H. (2014). Analytical demonstrations to assess residual bearing capacities of steel plate girder ends with stiffeners damaged by corrosion. *Structure and Infrastructure Engineering, 10*, 69–79.

Kim, I. T., & Itoh, Y. (2007). Accelerated exposure tests as evaluation tool for estimating life of organic coatings on steel bridges. *Corrosion Engineering, Science and Technology, 42*, 242–252.

Klinesmith, D. E., McCuen, R. H., & Albrecht, P. (2007). Effect of environmental conditions on corrosion rates. *Journal of Materials in Civil Engineering, 19*, 121–129.

Kuroda, S., Kawakita, J., & Takemoto, M. (2006). An 18-year exposure test of thermal-sprayed Zn, Al, and Zn–Al coatings in marine environment. *Corrosion, 62*, 635–647.

Landolfo, R., Cascini, L., & Portioli, F. (2010). Modeling of metal structure corrosion damage: A state of the art report. *Sustainability, 2*, 2163–2175.

MAINLINE. (2013a). *Deliverable 2.2: Degradation and intervention modelling techniques*, Work Package 2, MAINLINE (Maintenance, Renewal and Improvement of Rail Transport Infrastructure to reduce Economic and Environmental Impacts) Consortium. Retrieved from www.mainline-project.eu

MAINLINE. (2013b). *Deliverable 2.3: Time-variant performance profiles for life-cycle cost and life-cycle analysis*, Work Package 2, MAINLINE (Maintenance, Renewal and Improvement of Rail Transport Infrastructure to reduce Economic and Environmental Impacts) Consortium. Retrieved from www.mainline-project.eu

MAINLINE. (2013c). *Deliverable 5.4: Proposed methodology for a life cycle assessment tool (LCAT)*, Work Package 5, MAINLINE (Maintenance, Renewal and Improvement of Rail Transport Infrastructure to reduce Economic and Environmental Impacts) Consortium. Retrieved from www.mainline-project.eu

MAINLINE. (2014). *Deliverable 2.4: Field-validated performance profiles*. Work Package 2, MAINLINE (Maintenance, Renewal and Improvement of Rail Transport Infrastructure to reduce Economic and Environmental Impacts) Consortium. Retrieved from www.mainline-project.eu

Nguyen, T., Hubbard, J. B., & Pommersheim, J. M. (1996). Unified model for the degradation of organic coatings on steel in a neutral electrolyte. *Journal of Coatings Technology, 68*, 45–56.

NR. (2009a). *NR/L3/CIV/002: The use of protective coatings and sealants. Guidance Note*. Network Rail, UK.

NR. (2009b). *NR/L3/CIV/039: Level 3 – Specification for the assessment and certification of protective coatings and sealants*. Network Rail, UK.

NR. (2009c). *NR/L3/CIV/040: Level 3 – Specification for the use of protective coating systems*. Network Rail, UK.

Pommersheim, J. M., Nguyen, T., Zhang, Z., & Hubbard, J. B. (1994). Degradation of organic coatings on steel: Mathematical models and predictions. *Progress in Organic Coatings, 25*, 23–41.

Prucz, Z., & Kulicki, J. M. (1998). Accounting for effects of corrosion section loss in steel bridges. *Transportation Research Record: Journal of the Transportation Research Board, 1624*, 101–109.

Salas, O., de Tincon, O. T., Rojas, D., Tosaya, A., Romero, N., Sanchez, M., & Campos, W. (2012). Six-year evaluation of thermal-sprayed coating of Zn/Al in tropical marine environments. *International Journal of Corrosion, 2012*, Article ID 318279.

Sharifi, Y., & Paik, J. K. (2011). Ultimate strength reliability analysis of corroded steel-box girder bridges. *Thin-Walled Structures, 49*, 157–166.

Sørensen, P. A., Kiil, S., Dam-Johansen, K., & Weinell, C. E. (2009). Anticorrosive coatings: A review. *Journal of Coatings Technology and Research, 6*, 135–176.

Tamakoshi, T., Yoshida, Y., Sakai, Y., & Fukunaga S. (2006). Analysis of damage occurring in steel plate girder bridges on national roads in Japan. 22nd US–Japan Bridge Engineering Workshop, Seattle, WA.

Timoshenko, S. P. (1964). *Theory of plates and shells*. London: McGraw Hill.

van de Lindt, J. W., & Ahlborn, T. M. (2005). *Development of steel beam end deterioration guidelines* (Final Report). Houghton, MI: Michigan Tech, Research Report RC-1454.

Wardhana, K., & Hadipriono, F. C. (2003). Analysis of recent bridge failures in the United States. *Journal of Performance of Constructed Facilities, 17*, 144–150.

Zonta, D., Zandonini, R., & Bortot, F. (2007). A reliability-based bridge management concept. *Structure and Infrastructure Engineering, 3*, 215–235.

Simplified probabilistic model for maximum traffic load from weigh-in-motion data

Miriam Soriano, Joan R. Casas ⓘ and Michel Ghosn

ABSTRACT

This paper reviews the simplified procedure proposed by Ghosn and Sivakumar to model the maximum expected traffic load effect on highway bridges and illustrates the methodology using a set of Weigh-In-Motion (WIM) data collected on one site in the U.S.A. The paper compares different approaches for implementing the procedure and explores the effects of limitations in the site-specific data on the projected maximum live load effect for different bridge service lives. A sensitivity analysis is carried out to study changes in the final results due to variations in the parameters that define the characteristics of the WIM data and those used in the calculation of the maximum load effect. The procedure is also implemented on a set of WIM data collected in Slovenia to study the maximum load effect on existing Slovenian highway bridges and how the projected results compare to the values obtained using advanced simulation algorithms and those specified in the Eurocode of actions.

Introduction

Over the last two decades, highway agencies have recognised the importance of having automated data collection systems that can provide information on truck weights and truck traffic patterns for economic analysis, traffic management and various other purposes. To meet that goal, various types of Weigh-In-Motion (WIM) systems have been widely deployed to collect large quantities of unbiased truck data at normal highway speeds in an undetected manner to avoid truck driver's knowledge. WIM equipment currently in use can collect data on truck volumes, axle configurations, axle weights and truck arrival times. These WIM systems are based on different technologies with varying levels of accuracy, long-term performance and cost. Among available technologies, piezoelectric sensor-based WIM systems offer acceptable accuracy (usually ±15% for gross weights) at such low cost that their use has become quite widespread (Sivakumar, Ghosn, & Moses, 2011).

Although primarily used for traffic management purposes, WIM data have also been used to develop new live load models and assess the safety of new bridge designs and perform bridge ratings for existing bridges (Casas & Gomez, 2013; Crespo-Minguillón & Casas, 1997; Frangopol, Goble, & Tan, 1992; Fu & Tang, 1995; Ghosn & Moses, 1986; Ghosn, Moses, & Gobieski, 1986; Jacob, 1991; Moses, Ghosn, & Snyder, 1984; Nowak, 1999; Nowak & Nassif, 1992; O'Brien & Znidaric, 2001; O'Connor, Jacob, O'Brien, & Prat, 1998). Wiśniewski, Casas, and Ghosn (2012) propose the use of WIM data as one of the key factors in future developments of codes for advanced bridge assessment.

The use of WIM data for bridge evaluation requires a careful evaluation of the quantity and quality of available data.

For example, a study by Laman and Nowak (1997) shows that truck loads are strongly site specific and depend on traffic volume, local industry and law enforcement effort. A negative correlation was found between law enforcement effort and the occurrence of overloaded trucks. This observation may have significant impact on the safety of bridges in regions with low truck weight monitoring. Truck load growth trends have also been assessed utilising WIM data. For instance, a large-scale California study has established that truck volumes have increased over time even though gross vehicle weight spectra in the state have remained largely unchanged (Lu et al., 2002). Increases in traffic volumes may also lead to higher congestion and the bunching of heavy trucks over bridges increasing their risks.

The observed variations in truck weights and volumes led Ghosn and Frangopol (1996) to focus on the use of WIM data to define site-specific bridge loads. They highlighted the differences in estimated safety levels that result from applying site-specific values for bridge evaluation rather than national average load data. Several other studies have also used WIM traffic data to demonstrate how current code-specified load models could provide non-conservative estimates of actual loads experienced by some bridges (ARCHES D-08, 2009; Enright & O'Brien, 2013; Ghosn, Sivakumar, & Miao, 2013). Conversely, in other cases, WIM data revealed that many bridges are designed for traffic live loads that they will never experience during their expected service life (Casas, 1999). This fact is of relevant importance when facing the challenge of assessing the safety of existing bridges and in managing the limited resources available for rehabilitating and upgrading or ageing bridge infrastructure.

The examination of truck multiple presence on bridges has employed WIM data to simulate multi-lane traffic, and determining critical loading events and extreme load effects. It was generally noted that as the span length increases, the critical loading event is governed by an increasing number of trucks (Caprani, Grave, O'Brien, & O'Connor, 2002). Several studies indicated that traffic density should be a deciding factor in the development of multiple presence reduction factors or in assigning live load factors for bridge evaluation (Gindy & Nassif, 2006; Moses, 2001; Sivakumar et al., 2011).

WIM data have also been applied to the development of new fatigue models and the assessment of existing models (Casas, 2000; Crespo-Minguillón & Casas, 1998; Moses & Ghosn, 1985; Moses, Schilling, & Raju, 1987). Cyclic fatigue investigations reveal that fatigue load spectra are highly site specific and that code-specified load models often misrepresented actual fatigue damage even after accounting for the safety factors (Cohen, Fu, Dekelbab, & Moses, 2003; Grundy & Boully, 2004; Laman & Nowak, 1996; Moses et al., 1987; Wang, Liu, Huang, & Shahawy, 2005).

As discussed above, WIM data have many applications in bridge engineering. One important application is projecting the data to estimate the maximum traffic actions that bridges may be subjected to within a reference period. Several methods have been proposed to this end (O'Brien et al., 2015; Treacy, Brühwiler, & Caprani, 2014). However, most of these are based on complicated and cumbersome simulation processes that require excessive computational effort and specialised expertise making them difficult to implement in everyday engineering practice when evaluating the safety of specific bridges.

This paper illustrates the implementation of variations on a simplified procedure developed by Ghosn and Sivakumar (2010 and Sivakumar et al. 2011) to estimate the maximum expected load effect on a highway bridge. The procedure explains how site-specific WIM truck weight and traffic data can be used to obtain estimates of the maximum live load for the service life of a bridge using an easily implementable simple technique. Obtaining statistical information on the maximum live load expected in a bridge's design life or service period constitutes an important step in the reliability evaluation of the safety of the bridge's members and structural system. Also, the information assembled is necessary for calibrating the appropriate live load factors or for determining the appropriate return period that should be used while designing new bridges or evaluating the safety of existing ones (Jacob, 1991; Moses, 2001; Nowak, 1999).

The application of the simplified procedure is described in this paper using data collected at a WIM site in New York State. A sensitivity analysis is performed to identify the most critical parameters that control the projection of the expected maximum load and to study the effect of limitations in the WIM database. The validity of the procedure is verified by implementing it on a set of WIM data collected in Slovenia that had been previously analysed using more complex methods (ARCHES D-08, 2009; Enright & O'Brien, 2013).

Analysis of maximum load effect

Bridges must safely support the maximum traffic load expected within their service lives. This maximum load may be due to crossing over the bridge of a single heavy truck or a number of trucks simultaneously. The governing load depends on the axle spacing and weights of the trucks that cross in each lane of the bridge and the probability of simultaneous truck crossings. The probability of simultaneous crossings depends on the traffic characteristics including headway spacing and bridge length.

Ideally, such data should be available for an entire period equal to the service life of the bridge. However, despite the recent proliferation of WIM systems, collecting the necessary data to obtain the maximum load is not possible because decisions on the safety of a particular bridge should be made before exposing it to the most hazardous loads and because short-term data may not actually contain the worst possible loading scenarios. Also, data previously collected on one site may not necessarily represent future loads at the same or at different sites. For these reasons, the safety assessment of bridges must be performed using probabilistic methods based on a statistical projection of the maximum expected load from a set of data collected over a relatively short period of time.

The load effect of each truck on a particular bridge can be obtained by passing the truck through the appropriate influence line. Of particular importance for bridge designers and evaluators are the moment and shear effects at critical cross sections. The maximum loads on short- to medium-span bridges are governed by moving trucks rather than congested truck conditions (Caprani et al., 2002; Nowak, 1999). Because of the nature of truck traffic as well as the moment and shear influence lines of short- to medium-span bridges, the maximum load on multi-lane bridges with spans less than 60 m is governed by a single heavy truck, side-by-side trucks, or for continuous spans, two trucks following in the same or different lanes.

These observations render the modelling of the maximum load effects easy to implement using basic concepts of probability theory. In that vein, the ensemble of the load effects for each truck can be assembled into a histogram where each bin associated with a load effect x_i, gives the percentage of the load effects ranging between x_{li} and x_{ui}. The histogram value in each bin $H(x_i)$ would be related to the probability distribution of the truck effect $f_x(x)$ by:

$$H\left(x_i\right) = \int_{x_{li}}^{x_{ui}} f_x(x)\mathrm{d}x \tag{1}$$

The total load effect when two side-by-side (or following) trucks are simultaneously on a bridge is obtained from $S = x_1 + x_2$, where x_1 is the effect of the main truck and x_2 is the effect of the second truck. The explanations provided in this paper assume two-lane short- to medium-span bridges. However, the same formulation can be used for two trucks following each other in different spans of continuous bridges and the approach can be generalised for multi-lane bridges.

As shown in Figure 1, WIM data collected from several sites in New York State show that there is no correlation between the weights of trucks close to each other in the same lane or in adjacent lanes (Fiorillo, 2015). Hence, the probability density function of the effect of two side-by-side trucks $f_s(S)$ can be calculated using a convolution equation presented as:

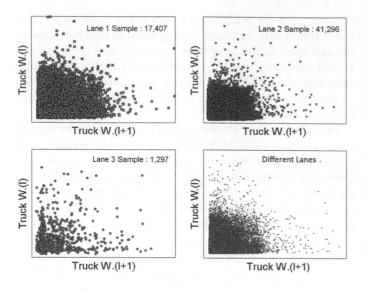

Figure 1. Correlation of consecutive trucks in lane 1, lane 2, lane 3 and in different lanes (Fiorillo, 2015).

$$f_s(S) = \int_{-\infty}^{+\infty} f_{x2}(S - x_1) f_{x1}(x_1) \, dx_1 \qquad (2)$$

where $f_s(...)$ is the probability distribution of the side-by-side effects, $f_{x1}(...)$ is the probability distribution of the effects of trucks in lane 1, $f_{x2}(...)$ is the probability distribution of the effects of trucks in lane 2. The cumulative distributions $F_s(S)$ of the single lane loading event or that of the side-by-side event can be obtained by integrating $f_{x1}(...)$ and $f_s(...)$ so that $F_s(S)$ denotes the probability that the load effect, s, is less or equal to a value equal to S.

A bridge structure should be designed to withstand the maximum load effect expected over the service life of the bridge. For example, the AASHTO LRFD (2014) specifies a design life of 75 years. In the case of the Eurocodes, the design life for bridges is 100 years. The LRFR bridge load rating from AASHTO MBE (2011) requires checking the capacity to resist the maximum load effects in a 5-year rating period, while bridges should be inspected every 2 years.

It is simply impossible to collect sufficiently large data-sets to determine the maximum load effect expected over 75 or 100 years of loading. Even collecting sufficient data for a 2-year inspection period would require several cycles of 2-year data and one is never assured that data collected in the past will actually represent future load spectra. Therefore, some form of statistical projection should be performed with available WIM data. The calculation procedure proposed by Sivakumar et al. (2011) uses the cumulative distribution function for individual loading events and then applies a statistical projection to obtain the information required for a 1-year, 2-year, 75-year or 100-year service or design life.

To find the cumulative distribution for the maximum loading event in a period of time, T, we have to start by estimating the number of loading events, N, that may occur during this period. These events are designated as $S_1, S_2 ..., S_N$. The maximum of these N events, call it $S_{max,N}$, is defined as:

$$S_{max,N} = \max(S_1, S_2, \ldots S_N) \qquad (3)$$

The cumulative probability distribution of $S_{max,N}$, $F_{s\,max\,N}(S)$, gives the probability that $S_{max,N}$ is less or equal to a value S. If $S_{max,N}$ is less than S, this implies that all the S_i are less than S. Hence, assuming that the loading events are independent (this can be justified based on Figure 1) but drawn from the same probability distribution, the probability that $S_{max,N} \leq S$ can be calculated from:

$$F_{s_{max}}(S) = \left[F_s(S)\right]^N \qquad (4)$$

The number of events N to be used in Equation (4) can be easily obtained from the WIM data for either the single lane loading cases or multiple truck events. It is interesting to note that for simple span bridges, several analyses using Equation (4) have shown that when the truck weight data include a large percentage of overloaded or permit trucks, the single truck event would govern the maximum loading of the bridge producing an $F_{s\,max\,N}(S)$ curve to the right of that of the side-by-side case; whereas for sites where all heavy trucks are within or close to the legal limits, the side-by-side case governs (Ghosn et al., 2013; Sivakumar et al., 2011).

The implementation of Equation (4) is straightforward only if the cumulative distribution function $F_s(S)$ is well defined with high precision in the upper tail end for large values of S. However, in practice, the use of Equation (4) is not directly possible for large values of N when the upper range of $F_s(S)$ is not well defined and it is necessary to execute some form of statistical projection of the tail end of $F_s(S)$.

To address the problem, researchers have proposed different methods to extrapolate the results from the limited number of WIM observations that can be collected in the field. For example, O'Brien and his colleagues (Enright & O'Brien, 2013; Enright, O'Brien, & Dempsey, 2010) have resorted to curve fitting and extrapolating the basic truck gross and axle weight histograms as well as the probability distribution of the headways. They used the fitted models in a Monte Carlo simulation program running for millions of cycles to obtain a cumulative distribution $F_s(S)$ with a very long extrapolated tail that can be either used directly into Equation (4) to find the probability distribution of the maximum load effect $F_{s\,max\,N}(S)$ or to find the load effect corresponding to a specified return period.

Other researchers (Moses, 2001; Nowak, 1999; Sivakumar et al., 2011) have opted for less demanding extrapolation approaches that are significantly more efficient and easier to implement than the Monte Carlo simulations advocated by Enright et al. (2010). The alternative projection techniques are significantly more efficient and easier to implement but still yield similar results to those of the Monte Carlo simulations. Specifically, Ghosn and Sivakumar (2010 and Sivakumar et al. 2011) proposed a systematic step-by-step approach that is easy to implement in engineering practice on a routine basis. While the proposed method does not differ much from other methods such as that followed by Nowak, Ghosn and Sivakumar removed much of the subjective decision-making steps that other researchers have used so that the extrapolation can be executed following simple rules that have been demonstrated to work well with several data-sets assembled from different WIM sites in the U.S.A.

In this paper, the authors review the extrapolation rules proposed by Ghosn and Sivakumar (2010) and verify their applicability to two sets of WIM data. The first set was collected on one site in New York State and the other in Slovenia. A few variations on the approach are also proposed to further simplify the procedure. A parametric analysis is performed to investigate the sensitivity of the results to the quality and quantity of WIM data available for executing the extrapolation.

Description of the model

To illustrate the procedure, one year of WIM data representing 531,445 trucks was obtained for the North Bound direction of the I-81 Highway in upstate New York for analysis. Each truck was sent through the influence line for the moment at the mid-span of a 20-m bridge. The moments were then normalised as a function of the moment effect of the HL-93 load in the AASHTO (2014) specifications and assembled into a histogram as shown in Figure 2 for both lanes of traffic (drive and passing lane). As shown in Figure 2, the probability distribution of the single loading event does not follow any known probability distribution type.

Although it is possible to represent such histograms by composite distributions that simulate the histograms' multi-modal nature, such intricate analysis is not necessary when the goal is to estimate the maximum load effect which is governed by the heaviest trucks that affect the tail end of the load histogram and the maximum load effects. It is for this reason that Ghosn and Sivakumar (2010) focused on modelling the tail end of the histogram, observing that the tail end approaches that of a normal distribution.

This is illustrated in Figure 3 which plots the load effect data for the drive lane of the I-88 WIM site on normal probability scale which shows that the upper 5% of the data falls along a straight line with a regression coefficient R^2 = .997. This means that the upper 5% matches that of a hypothetical normal distribution with a mean value μ_x = .023 and a standard deviation value σ_x = .33. A similar result is obtained for the case of the passing lane. After analysing many WIM data-sets collected in widely different regions of the U.S.A., Sivakumar et al. (2011) observed the same trend in all the data-sets they analysed. The upper 5% of the data showed high goodness-of-fit statistics and coefficients of determination when fitted to tail ends of normal distributions. The 5% criterion for the fit was decided upon as a result of these evaluations.

The observations made with U.S.A. WIM data-sets with respect to fitting the upper 5% to normal distributions seem to be also valid for data collected in Europe as shown in Figure 4 that plots normalised moment effect on a 30-m simple span from WIM truck data recorded on a Slovenian highway (ARCHES D-08). In this case, the normal distribution fit was executed directly on the tail of the histogram, rather than on normal probability scales, by a least-square minimisation error method as shown in Figure 4.

For high numbers of loading events, N, the application of Equation (4) is controlled by the tail end of $F_s(S)$ near the upper 5% of the data. Since this region matches that of a normally distributed random variable, a closed form representation of the mean and standard deviation of S_{max} can be directly obtained from the mean and standard deviation of a single event μ_x and σ_x. The projection is based on extreme value principles, which state that if the parent distribution of a variable, S, follows a normal probability function, the distribution of the maximum value S_{max}

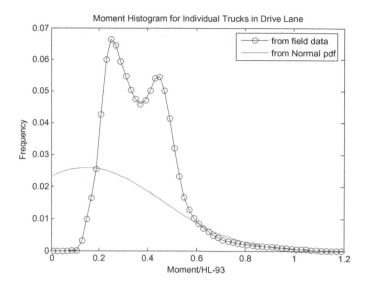

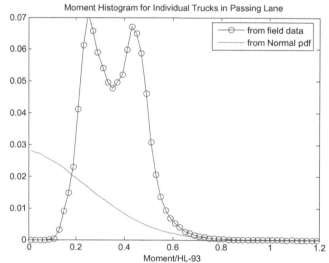

Figure 2. Normalised moment histogram for trucks in drive lane (up) and in passing lane (bottom) of I-81 NB.

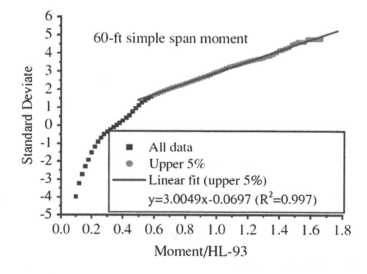

Figure 3. Normal probability plot of moment effect on a 20-m bridge using I-81 data (drive lane).

approaches a Gumbel distribution as N increases with an inverse dispersion coefficient α_N given by (Ang & Tang, 2006):

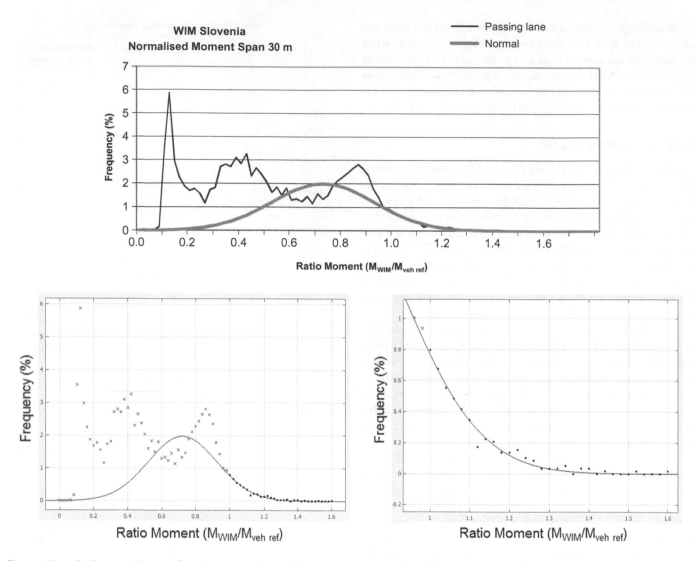

Figure 4. Normalised moment histogram for a 30-m span and a detail of regression fit of upper tail with Normal distribution (Slovenian traffic, passing lane).

$$\alpha_N = \frac{\sqrt{2\ln(N)}}{\sigma_X} \tag{5}$$

and a most probable value given by:

$$u_N = \mu_X + \sigma_X\left(\sqrt{2\ln(N)} - \frac{\ln(\ln(N)) + \ln(4\pi)}{2\sqrt{2\ln(N)}}\right) \tag{6}$$

This will lead to a mean of the maximum load effect:

$$L_{\max} = \mu_{\max} = u_N + \frac{0.577216}{\alpha_N} \tag{7}$$

and a standard deviation:

$$\sigma_{\max} = \frac{\pi}{\sqrt{6}\alpha_N} \tag{8}$$

Projection approach

While the statistical projection method used by Ghosn and Sivakumar (2010) is based on extreme value theory, other researchers (Nowak, 1999; Moses, 2001; Enright, et al., 2010)

used the return period concept whereby the plotted cumulative distribution data are fitted through a curve that is extrapolated to obtain the probability of exceedance corresponding to the number of events in a given period, T_r. That is, if a 75-year period is selected for a bridge that is expected to be crossed by 2000 trucks/day, the value of the load corresponding to a probability of exceedance $F_s(S) = 1/N$, where $N = 2000 \times 365 \times T_r$ is used to estimate the maximum load expected in a 75-year service period. The return period approach is consistent with the formulation adopted in the Eurocodes that stipulate a 1000-year return period for lives loads on bridges.

Another projection approach consists of taking the maximum load observed within a basic unit of time, estimating the probability distribution of the maximum value and then applying Equation (4) where N in this case is the ratio between the basic time unit and the service period (Sivakumar et al., 2011). Other researchers such as Fu and You (2011) used that approach and assembled a histogram of the maximum monthly load, fitted that histogram into a Gumbel probability distribution and raised the cumulative distribution of the Gumbel to a power $N = 12 \times 75$ to obtain the distribution of the maximum load in 75 years. As observed in Sivakumar et al. (2011), the maximum monthly fit

Table 1. Effect of tail end model on mean and standard deviation of maximum load effect.

Projection period	Drive lane loading						Side-by-side loading					
	1 week		2 years		75 years		1 week		2 years		75 years	
	M	S.D.	*M*	S.D.	*M*	S.D.	*M*	S.D.	*M*	S.D.	*M*	S.D.
Normal fit to tail end of WIM hist.	1.36	0.091	1.67	.074	1.87	.066	1.21	.171	1.75	.122	2.08	.104
Return period approach	1.31	N.A.	1.63	N.A.	1.84	N.A.	1.12	N.A.	1.69	N.A.	2.03	N.A.
Gumbel fit to max. weekly load	1.37	.068	1.62	.068	1.81	.068	1.44	.099	1.80	.099	2.08	.099
Lognormal fit to tail end of WIM hist.	.95	.122	1.22	.099	1.48	.089	1.50	.091	1.77	.067	1.91	.058
Gumbel fit to tail end of WIM hist.	1.51	.089	1.78	.073	1.97	.065	2.30	.233	2.96	.172	3.30	.141

Notes. *M* = mean; S.D. = standard deviation.

approach will yield similar results to those of the model presented here assuming that the data are stationary and that the truck traffic pattern and truck weights are essentially similar throughout the year. However, by ignoring much of the data and concentrating only on a single maximum value observed over a relatively large time period, the approach may yield inaccurate results if, for example, a one-month period is exposed to a large number of high loads which will all be ignored except for a single one of these.

In addition, the maximum monthly fit approach will require very long periods of data collection in order to collect sufficient numbers of monthly maxima to obtain a good fit for the probability distribution function. In order to increase the data-set, one could resort to fitting the maxima from shorter periods such as the weekly maxima or daily maxima. However, in these cases, the goodness of the Gumbel fit distribution is reduced. In fact, the data analysed as part of NCHRP 12-76 (Sivakumar et al., 2011) demonstrate that even though the maximum load over an extended period of time may approach a Gumbel probability distribution function, the same is not necessarily true for the maximum load observed over a one-week or one-month period. This has led some researchers to use generalised forms of the extreme value distribution fitted to the one-day maxima (O'Brien et al., 2015). It should be noted, however, that fitting the maxima of short durations will emphasise the contributions of the lower end of the data-set rather than those in the tail end which do contribute the most to the service life maxima. Using the wrong distribution has an important influence on the final result.

A set of analyses are performed based on the I-88 data to compare the results using different fitting approaches and study the effect of errors in the distribution type. Table 1 compares the results obtained by Sivakumar et al. (2011) for the maximum projected normalised moment effect for the 20-m simple span bridge using: (1) a Normal distribution fit to the tail end of the WIM data, (2) a Gumbel distribution fit to the tail end of the maximum one-week load as extracted from the WIM data, (3) the projected normal distribution load for a return period corresponding to the service life, as proposed by Nowak and Nassif (1992) and Moses (2001). In this case, the maximum load effect is approximated by $L_{max} = \mu_x + \sigma_x \Phi^{-1}(1/N)$, where μ_x and σ_x are the mean and standard deviation of the equivalent normal distribution modelling the upper 5% of the tail, Φ^{-1} is the inverse of the cumulative normal distribution function and N is the number of events expected in the service period, (4) a Gumbel distribution fit to the tail end of the histogram, similar to what is proposed here but using a Gumbel distribution instead of a Normal distribution and (5) a Lognormal distribution fit to the tail end of

the load histogram, similar to what is proposed here but using a Lognormal distribution instead of a Normal distribution. The following observations are made:

- The results of the normal fit, the weekly Gumbel fit and the normal return period give similar mean of maximum load value as the projection period increases.
- The use of the return period approach cannot provide the standard deviation of the maximum load value.
- The Gumbel fit to the weekly load requires the elimination of a large amount of the collected data and fitting only the tail end of the distribution of the maximum weekly load which reduces the confidence level in the projected mean value.
- Unless justified by a statistical evaluation of the data through appropriate goodness-of-fit tests, using a lognormal or a Gumbel fit to the tail end would lead to large errors in the results.

The fit of the WIM histogram's tail into a normal distribution helps simplify the multi-presence problem. For example, for side-by-side trucks in two-lane bridges, the convolution in Equation (2) reduces to a normal distribution with a mean value equal to the sum of the mean of trucks' load effects in the drive lane to that of the trucks' in the passing lane. The standard deviation of the combined effect is the square root of the sum of the squares if the loads in the two lanes are independent which is justified based on the data plotted in Figure 1.

As mentioned earlier, another issue is related to the technique used to obtain the statistics of the fitted tail end. For example, Sivakumar et al. (2011), following Nowak (1999) used a linear regression fit to the tail end of the cumulative distribution plotted on normal probability scale to obtain the statistics of the normal distribution function that matches the tail end of the actual histogram. Another approach would directly use the frequency histogram to obtain statistics of the Normal distribution.

Experimentations with these two approaches have demonstrated that in certain cases, the two different fitting techniques may lead to seemingly different results as shown in Figure 5, that shows the goodness of the fit when it is executed on the cumulative distribution as compared to the goodness of the fit when the statistics are extracted to match the tail end of the frequency histogram for the same data plotted on a normal cumulative scale and on a frequency histogram.

However, the effect of these errors on the projected maximum load is relatively small. For example, the mean and standard deviations of the normal distribution fitted to the tail end of the cumulative of the normalised moments of the 20-m

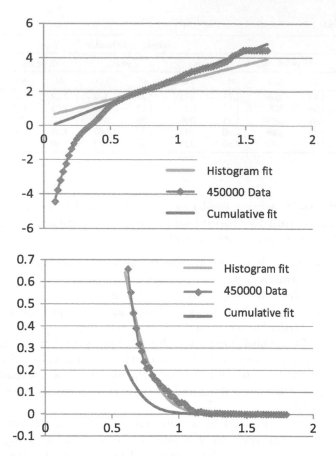

Figure 5. Illustration of differences in results of fits to cumulative function or histogram.

bridge using a randomly selected 450,000 samples produce on the average $\mu_x = .046$ and a standard deviation value $\sigma_x = .29$, while the fit to the tail end of the frequency histogram would yield a mean $\mu_x = .25$ and a standard deviation value $\sigma_x = .27$. The projected maximum 75-year load in the first case would be $L_{max} = \mu_{max} = 1.81$ vs. $L_{max} = \mu_{max} = 1.89$ in the second case. This demonstrates that errors in the mean value for one event are possible because this value happens to be close to zero but the errors' effect on the projected service life maximum is within an acceptable range of less than 5%.

Sensitivity analysis

To verify the robustness of the proposed method for obtaining the maximum load effect, an analysis is carried out to check the sensitivity of the results obtained using the I-88 data to changes in the main parameters of the model.

Effect of tail end of truck load effect histogram

The analysis of the WIM data histogram used in this study shows that the tail end of the data in the drive lane matches the tail end of a Normal distribution with a mean $\mu_{x1} = .1472$ and a standard deviation $\sigma_{x1} = .3074$. Similarly, the tail end of the histogram for the passing lane shows that it matches that of a Normal distribution with mean $\mu_{x2} = -.0417$ and a standard deviation $\sigma_{x2} = .2797$. In this section, it is examined how a change in these parameters will affect the values of the expected maximum load effect, L_{max}, of Equation (7) for the two-year and 75-year time periods. A simulation is executed to produce a histogram having a tail end that reflects the changes in these mean and standard deviation values.

In a first step, the mean values of the Normal distributions of the drive lane and the passing lane are changed simultaneously by the same ratio, while the standard deviations are kept at their original values. In the second step, the standard deviations of the drive lane and passing lanes are changed by the same ratio, while the mean values are kept at their original values. The convolution approach with the Gumbel projection is performed and the results of L_{max} obtained from Equation (7) are provided in Table 2. The third and fourth columns give the mean and standard deviations of L_{max} for the two-year service period. The fourth and fifth columns give the mean and standard deviations of L_{max} for the 75-year design life. The results in Table 2 illustrate that the mean value of L_{max} is not significantly affected when the mean value of the normal distribution is changed. Specifically, a change of 20% in the mean value of the equivalent normal distribution that matches the tail end of the load effect histogram leads to a change of less than 1.5% in the expected maximum load effect L_{max}. On the other hand, the table shows that L_{max} is more sensitive to changes in the standard deviation of the equivalent normal distribution. For example, when the standard deviation is changed by a factor of 10%, the expected maximum load effect changes by as much as 7.6%.

Table 2. Sensitivity of expected maximum load effect L_{max} to changes in statistical properties of histogram's tail end.

		Mean of tail (μ_x=original mean)	Stand. Dev. of tail (σ_x=original)	L_{max} for 2-year max.	L_{max} for 75-year max
Side-by-side loading	Effect of change in μ_x	$.80\mu_x$	$1.00\sigma_x$	1.772	2.021
		$.90\mu_x$	$1.00\sigma_x$	1.780	2.030
		$1.00\mu_x$	$1.00\sigma_x$	1.789	2.038
		$1.10\mu_x$	$1.00\sigma_x$	1.797	2.047
		$1.20\mu_x$	$1.00\sigma_x$	1.805	2.055
	Effect of change in σ_x	$1.00\mu_x$	$.90\sigma_x$	1.678	1.889
		$1.00\mu_x$	$.95\sigma_x$	1.734	1.962
		$1.00\mu_x$	$1.00\sigma_x$	1.789	2.038
		$1.00\mu_x$	$1.05\sigma_x$	1.845	2.115
		$1.00\mu_x$	$1.10\sigma_x$	1.902	2.193
Drive lane loading	Effect of change in μ_x	$.80\mu_x$	$1.00\sigma_x$	1.570	1.759
		$.90\mu_x$	$1.00\sigma_x$	1.582	1.771
		$1.00\mu_x$	$1.00\sigma_x$	1.593	1.784
		$1.10\mu_x$	$1.00\sigma_x$	1.604	1.796
		$1.20\mu_x$	$1.00\sigma_x$	1.616	1.809
	Effect of change in σ_x	$1.00\mu_x$	$.90\sigma_x$	1.482	1.652
		$1.00\mu_x$	$.95\sigma_x$	1.539	1.719
		$1.00\mu_x$	$1.00\sigma_x$	1.593	1.784
		$1.00\mu_x$	$1.05\sigma_x$	1.644	1.844
		$1.00\mu_x$	$1.10\sigma_x$	1.692	1.901

Table 3. Effect of reduced data on tail end representation and maximum load (fit to the histogram).

WIM data samples	Approx. monitoring period	Mean μ_{event} min.	Mean μ_{event} average	St. dev σ_{event} min.	St.dev σ_{event} average	Mean μ_{max} min.	Mean μ_{max} average	St. dev σ_{max} min.	St. dev σ_{max} average
531,445	12 months	–	.2412	–	.27	–	1.8712	–	.058
450,000	10 months	.237500	.24583	.267400	.272115	1.865516	1.8886	.0597490	.058455
360,000	8 months	.238300	.24583	.265200	.271860	1.855234	1.8871	.0595122	.058396
270,000	6 months	.238500	.24826	.261600	.270370	1.838601	1.8805	.0598130	.058073
180,000	4 months	.235400	.24785	.259000	.272515	1.826404	1.8930	.0606294	.058509
90,000	2 months	.234400	.24817	.259300	.269860	1.829415	1.8773	.0601353	.057941
45,000	1 month	.210900	.25045	.242900	.269740	1.746707	1.8789	.0652916	.057816
21,000	2 weeks	.166700	.24139	.243700	.275505	1.750837	1.9046	.0704693	.058920

Table 4. Effect of reduced data on tail end representation and maximum load (fit to the cumulative).

WIM data samples	Approx. monitoring period	Mean μ_{event} min.	Mean μ_{event} average	St. dev σ_{event} min.	St. dev σ_{event} average	Mean μ_{max} min.	Mean μ_{max} average	St. dev σ_{max} min.	St. dev average
531,445	12 months	–	.0548	–	.2867	–	1.7850	–	.08249
450,000	10 months	−.05447	.046348	.247783	.291350	1.577757	1.8053	.08197107	.061750
360,000	8 months	.017614	.054057	.183361	.286113	1.281238	1.7813	.07148727	.060354
270,000	6 months	.006473	.041892	.256334	.300540	1.622054	1.8563	.07540447	.064600
180,000	4 months	.001705	.031759	.230690	.310294	1.417799	1.9050	.07619192	.065878
90,000	2 months	.004071	.050494	.164885	.294006	1.184273	1.8254	.07588406	.061330
45,000	1 month	.011264	.037795	.211249	.307120	1.378060	1.8919	.7550530	.064649
21,000	2 weeks	.007574	.051117	.217588	.305055	1.415754	1.8928	.07648175	.064489

The analysis performed in this section was based on creating simulated data in the upper 5% of the truck load effect histogram. Because of the simulated data, the results for the base case when the mean is $1.0\mu_x$ and the standard deviation is $1.0\sigma_x$ are slightly different than those obtained from the original histogram. Also, the analysis assumes that the tail end matches that of a normal distribution. The analysis performed in this section demonstrates the importance of the variation in the weights of the heaviest trucks on the prediction of the maximum load effect. Therefore, the data collection process should be well planned to ensure good confidence levels in the standard deviation of the heaviest trucks that are likely to cross the bridge site.

Effect of shorter term data

Sivakumar et al. (2011) recommend that at least 1 year of WIM data be collected for use in projecting the maximum load effect for the purpose of evaluating the safety of bridges. This recommendation is made in order to ensure that sufficient numbers of samples are available to give an accurate representation of the tail end of the histogram and also to account for any seasonal changes in the truck traffic pattern and gross weights.

When evaluating the safety of existing bridges, it is often not possible to assemble an entire year's data. This is because very few bridges are fitted with permanent WIM systems that can provide long-term data when needed. Instead, it is most common to use portable WIM systems that can be only deployed for short periods of time. To study the effect of shorter term data, a simulation is performed to study the effect of the reduction in the numbers of samples. This is executed by taking a random number of N samples from the WIM data for various values of N representing the equivalent of 10, 8, 6, 4, 2 and 1 months and two weeks of data.

The results of the smaller sample size on the mean and standard deviation of the hypothetical normal distribution representing the tail end of the single event and the projected mean and standard deviation of the maximum load effect are provided in Tables 3 and 4. Of course, the analysis presented assumes that the bridge loading process is stationary where the parent distribution does not change with time neither in the short nor in the long terms. It is also understood that because of the random nature of the problem, different simulations with the same number of samples may produce different results. Therefore, Tables 3 and 4 include the minimum as well as the average from 20 Monte Carlo runs for each sampling period. The minimum value is provided to give an approximation to the maximum error that could be obtained if a single set of data is used as compared to the average from 20 simulations.

Although the results in Tables 3 and 4 show large variations between the minimum values and the average values, no decipherable trend is observed in the average 75-year maximum obtained from 20 simulations, when the sampling period is changed. This indicates that what is important is to ensure that the sampling period should be targeted to catch representative data in the tail end of histogram rather than on the quantity of data. Certainly, taking data over the longest possible sampling period would minimise the chances of missing the heaviest trucks. Based on this observation, it is recommended that the deployment of WIM systems be carefully planned with input from traffic engineers to collect data over the longest period of time possible and ensure that the data collection coincides with periods of high truck traffic volumes and when truck with heavy cargo are in operation.

Effect of inaccuracy in WIM data

Another potential source of errors in the projected maximum load is the lack of accuracy of the WIM system used to collect the truck data. Any truck measurement system has inherent inaccuracies. Although a properly calibrated system will on the average produce accurate axle spacing and axle weights, some random errors are normally observed around the exact value for the best calibrated systems. Calibrated WIM systems have been found to produce errors with standard deviations ranging from around +/−5% up to +/−15% of the gross vehicle weights.

For the purposes of this study, we will assume that a similar range of errors is obtained on the load effect. This is done

Table 5. Effect of WIM measurement errors.

Stand dev of error (%)	Monitoring period	Mean μ_{event}	Stand. dev σ_{event}	Mean μ_{max}	Stand. dev σ_{max}
Histogram					
0	12 months	.2412	.27	1.8712	.058
0	10 months	.2458	.2721	1.8886	.0586
5	10 months	.2856	.2429	1.7518	.0521
15	10 months	.3535	.2043	1.5869	.0439
0	2 weeks	.2414	.2755	1.9046	.0589
5	2 weeks	.2830	.2446	1.7597	.0524
15	2 weeks	.3566	.2010	1.5700	.0431
Cumulative					
0	12 months	.0548	.2867	1.785	.0825
0	10 months	.0463	.2914	1.8053	.0618
5	10 months	.0064	.3601	2.1672	.0772
15	10 months	.1214	.3187	2.0452	.0680
0	2 weeks	.0511	.3051	1.8928	.0645
5	2 weeks	.0016	.3557	2.1492	.0759
15	2 weeks	.1135	.3197	2.0435	.0666

by adding a random error to each truck response. A sensitivity analysis is then performed to observe the effect of random errors on the projected maximum load, L_{max} calculated using Equation (7). The results for the projected maximum load effect assuming a 12-month monitoring period are compared to the maximum load obtained from a 10-month and two-week monitoring periods in Table 5. In this case also, the values in Table 5 are obtained based on 20 Monte Carlo runs to account for the random nature of the data sampling process. Table 5 shows how the results are more sensitive to measurement errors when the normal fit is executed based on the tail end of the histogram, while the projection of the maximum load effect obtained by fitting the cumulative probability plot on normal scale is less sensitive. Reducing the monitoring period from 10 months to 2 weeks does not significantly affect the results or amplify the effect of the errors.

Effect of truck traffic intensity

The number of loading events expected within a return period has been defined as N in Equations (4)–(6). Although this is an important parameter, previous studies have demonstrated that the final results asymptotically approach an upper limit as the number of events N increases. This would be especially true when the service life period exceeds 50 years as required for estimating the maximum load effects when designing new bridges. The object of this paragraph is to study how the number of loading events affects the results of L_{max}. The analysis performed in this section assumes that the number of loading events per day is changed from the 2000 events obtained from the WIM data while keeping the percentage of the trucks that are side-by-side, P_{sxs}, constant at .95% as obtained from I-81 NB WIM data. Similarly, the raw WIM load effect histograms for the trucks in the drive lane and those in the passing lane are kept the same. The range of values used for the number of loading events is varied from 500/day to 6000/day as shown in Tables 6 and 7 for the 75-year and 2-year projections.

The results of Tables 6 and 7 confirm that a decrease in the number of events by a factor of 4 (from 2000 to 500) leads to a reduction in the expected maximum 75-year load effect for the loading in the drive lane by −6.59%, while increasing the number of events by a factor of 3 (from 2000 to 6000) leads to an increase in the expected maximum load effect by +3.26%.

Table 6. Summary of results from I-81 NB site for different ADTT: 75-year period.

			Simplified Gumbel
One lane loading	Mean 75 years	$N = 500$/day	1.794
		$N = 1000$/day	1.832
		$N = 2000$/day	1.870
		$N = 3000$/day	1.891
		$N = 4000$/day	1.907
		$N = 6000$/day	1.928
	Standard deviation 75 years	$N = 500$/day	.069
		$N = 1000$/day	.068
		$N = 2000$/day	.066
		$N = 3000$/day	.066
		$N = 4000$/day	.065
		$N = 6000$/day	.064
Side-by-side loading	Mean 75 years	$N = 500$/day	1.963
		$N = 1000$/day	2.024
		$N = 2000$/day	2.083
		$N = 3000$/day	2.117
		$N = 4000$/day	2.141
		$N = 6000$/day	2.174
	Standard deviation 75 years	$N = 500$/day	.110
		$N = 1000$/day	.107
		$N = 2000$/day	.104
		$N = 3000$/day	.102
		$N = 4000$/day	.101
		$N = 6000$/day	.100

This demonstrates that even large errors in the determination of the number of loading events and in the estimation of the Average Daily Truck Traffic (ADTT) would produce only marginal errors in the estimated maximum load effect. This is due to the asymptotic nature of the problem whereby the results of Equation (7) would converge as the number of events increases. It is noted that the results provided in Tables 6 and 7 assume that the percentage of events that are due to side-by-side trucks remains constant. The analysis of WIM data has shown that the percentage of side-by-side cases changes as the ADTT at the site changes. For instance, Figure 6 shows a strong correlation between multi-presence and ADTT for the New York traffic data (Sivakumar et al., 2011).

Effect of passing rates

A study of truck multiple-presence at 25 sites across New Jersey over a period of 11 years has also provided valuable data on the relationship between truck volume and truck multiple presence (Gindy & Nassif, 2006). Obtaining reliable multiple-presence

Table 7. Summary of results from I-81 NB site for different ADTT: 2-year period.

			Simplified Gumbel
One lane loading	Mean 2 years	$N = 500$/day	1.579
		$N = 1000$/day	1.623
		$N = 2000$/day	1.665
		$N = 3000$/day	1.689
		$N = 4000$/day	1.706
		$N = 6000$/day	1.730
	Standard deviation 2 years	$N = 500$/day	.078
		$N = 1000$/day	.076
		$N = 2000$/day	.074
		$N = 3000$/day	.073
		$N = 4000$/day	.073
		$N = 6000$/day	.072
Side-by-side loading	Mean 2 years	$N = 500$/day	1.609
		$N = 1000$/day	1.683
		$N = 2000$/day	1.753
		$N = 3000$/day	1.793
		$N = 4000$/day	1.820
		$N = 6000$/day	1.859
	Standard deviation 2 years	$N = 500$/day	.132
		$N = 1000$/day	.127
		$N = 2000$/day	.122
		$N = 3000$/day	.120
		$N = 4000$/day	.118
		$N = 6000$/day	.116

statistics requires large quantities of continuous WIM data with refined time stamps, which may not be available at every site. For this reason, it is important to study the effect of changes in the number of side-by-side events on the results of the maximum load effects. The effect of passing rates as expressed in terms of the per cent of side-by-side cases, P_{sxs}, on the values of the expected maximum load effects, L_{max}, is similar to the effect of changes in the number of events or ADTT in the sense that this will only affect the number of load repetitions N of Equations (4) and (7) for the case of side-by-side trucks.

For example, if the number of events per day is kept at 2000 and assuming that 6% of these events (instead of the actually measured by WIM and previously used .95%) are side-by-side, then the total number of side-by-side cases in one day becomes 120 instead of the original 19. If the number of single truck events in the drive and passing lanes remains proportional to the original distribution, then one could expect that 84.04% of the events will be single trucks in the drive lane and 9.96% will be single trucks in the passing lane. Thus, the number of side-by-side events becomes 120 and the number of daily events in the drive lane is 1800 (2000 × 84.04% + 120). Notice how the 1800 number of events in the drive lane remains very close to the original 1790 obtained when .95% of the events are side by side.

This indicates that the projection of the maximum load effect for the single lane cases will not be significantly affected by a change in the percentage of side-by-side trucks and that only the side-by-side cases will be affected. Even so, the changes in the maximum load effect, L_{max}, will be relatively small due to the asymptotic nature of the problem. This fact is illustrated in Table 8 which shows how little change is observed in the expected 75-year and 2-year maximum load effects with the percentage of side-by-side cases for different ADTT. For example, a change in the percentage of side-by-side events from .5 to 6% or by a factor of 12 (1200%) for a site with ADTT = 2000 results in a change in the expected load effect, L_{max}, for the 2-year return period of only 11%. Similar changes are observed for all the ADTT cases considered. These results indicate that even if the WIM data may contain large errors in the side-by-side count, the relative effect of these errors on the final results will not be significant.

The results from Tables 6–8 demonstrate that as long as the errors in estimating the ADTT and the per cent of side-by-side events remain within a factor of 3, the error in the expected maximum live load effect, L_{max}, will remain within 5% or less.

Discussion of the results

As observed from the sensitivity analysis, the projection of the maximum load effect for short- to medium-span bridges for 2-year and 75-year projection periods is primarily controlled by three factors: (1) the number of loading events expected within the projection period, (2) the percentage of the loading events that are closely spaced and, in particular, those due to

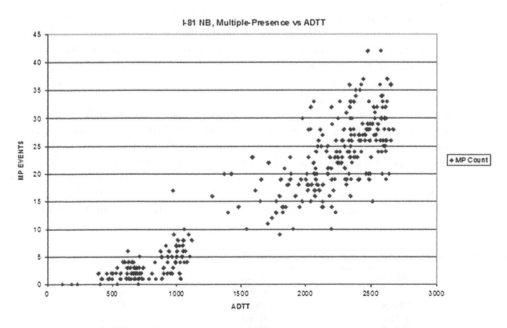

Figure 6. Side-by-side events vs. ADTT I-81 NB New York (adapted from Sivakumar et al., 2011).

Table 8. Effect of changes in side-by-side percentage on expected maximum load effect side-by-side cases only.

ADTT	Side-by-side percentage P_{sxs} (%)	Mean 2-year maximum load effect	Mean 75-year maximum load effect
2000	.5	1.749	2.029
	1	1.802	2.082
	2	1.856	2.136
	4	1.909	2.189
	6	1.941	2.220
4000	.5	1.802	2.082
	1	1.856	2.136
	2	1.909	2.189
	4	1.963	2.243
	6	1.994	2.274
6000	.5	1.834	2.113
	1	1.887	2.167
	2	1.941	2.220
	4	1.994	2.274
	6	2.026	2.305

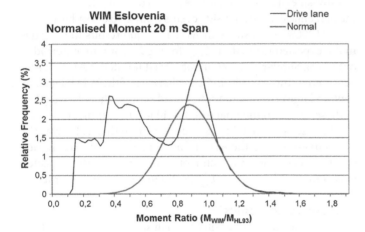

Figure 7. Moment effect of Slovenian traffic data on a 20-m span bridge (drive lane).

side-by-side trucks, (3) the trucks that fall within the tail end of the load effect histogram. The sensitivity analysis has demonstrated that a change in the number of events on the order of +/−300% leads to minor errors in the estimated maximum load effect of about +/−5%.

Similar results are observed for errors in the estimated percentage of side-by-side trucks. As shown, the most important parameters are those that describe the shape of the tail end of the histogram. In principle, the simplified analysis approach can handle any shape for the tail end of the load histogram assuming that the WIM data provide a full description of the tail. However, in reality, it is impossible to obtain 75-year worth of data to obtain a full description of the tail end of the histogram. Hence, the method must rely on some assumptions and approximations to the tail end of the histogram. Therefore, extreme care must be taken during the WIM data collection process to ensure that the

data are collected over long periods of time and is representative of the trucks with heaviest cargos.

Accuracy of the proposed model

In addition to the sensitivity analyses described earlier and the comparisons made for the different fitting techniques, the accuracy of the proposed model has also been compared with a simulation-based approach used by other researchers to estimate the maximum traffic load effect on highway bridges. A different WIM data-set from that used in the previous analysis is also considered. The two models were applied on a set of WIM data collected in Slovenia (Znidaric, Lavric, & Kalin, 2010) over a period of 58 days uninterruptedly. The total number of measured trucks was 147,752 and the ADTT was 3293. Figure 7 shows the moment effect of the trucks that crossed the drive lane of a bridge with a span of 20 m similar to the instrumented one (Casas, Ghosn, & Soriano, 2012). The figure also shows the normal distribution fit to the tail end of the moment histogram.

Table 9 compares the results obtained by various researchers for the maximum load effect for side-by-side trucks with the same WIM database for a simply supported 25-m bridge span. The researchers utilised the following approaches to project the load effects to obtain the maximum load in different service periods (kNm).

- Using the convolution of Equation (2) with the original WIM data.
- Using the normal fit to the upper tail end of the original histogram as proposed here.
- Large-scale Monte Carlo simulation taking into consideration, the weights of the trucks, the trucks' axle configurations, histograms of multiple presence and headway data as performed by Enright et al. (2010).
- A variation on the NCHRP method as performed by Znidaric (ARCHES D-08, 2009; Znidaric, Kreslin, Lavric, & Kalin, 2012).

Table 9 demonstrates that all four approaches lead to very similar results with a difference less than 3% which verifies that the approach proposed here is most applicable for engineering practice because of its simplicity and ease of implementation.

The projection of the results to estimate the maximum load effect over the bridge's service period requires as input the number of events in the service period N in Equations (4)–(6). After many simulations and comparisons, Znidaric et al. (2012) proposed to use the time of arrival of trucks as registered in the WIM data file and count the number of events that the trucks fall within the middle 50% of the bridge's influence line to estimate the number of simultaneous crossings of trucks. The approach, which depends on the speed of trucks, is illustrated in Figure 8. Table 10 gives

Table 9. Comparison of maximum load effect with different methods of analysis and different service periods (Slovenian traffic data). (kNm)

	WIM data	Znidaric	Normal fit	Enright
1 year	4676.1	4698.7	4800.2	−
5 years	5053.8	5062.2	5117.0	−
50 years	5510.5	5519.3	5530.1	5646.5
75 years	5586.9	5594.4	5598.7	5750.5

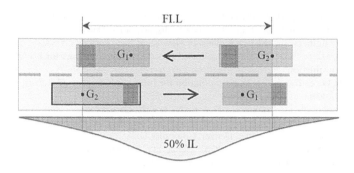

Figure 8. Modelling the number of simultaneous multiple crossings of trucks.

Table 10. Average number of two-truck events per day.

	20 m	30 m	40 m	50 m
80 km/hr	19	25	64	96
60 km/hr	23	64	101	148
40 km/hr	64	110	223	301

the average number of two trucks simultaneously on the bridge expected for different span lengths and traffic speeds.

The analysis of the Slovenian WIM data for the maximum moment on different span lengths and different service periods is presented in Table 11. From Table 11, it is seen that the maximum moment is obtained for the lowest speed, which produces the highest average number of two-truck events per day. This demonstrates that the side-by-side truck effect is the most critical for simple span bridges in the range of 20–50 m.

Applicability of the proposed model

Table 12 gives the load effect for a 1000-year return period obtained from the analysis with the Slovenia WIM data for different span lengths. The 1000-year is the return period recommended in the Eurocode for finding the characteristic value of traffic load effect. The values in Table 12 are obtained from the mean values in Table 11 assuming a coefficient of variation of the traffic action equal to 20%. The higher COV used in this case when compared to those obtained from the projections of the load effect accounts for site-to-site variability as well as analysis uncertainties. The listed 1000-year values also include a 1.10 dynamic amplification factor (Ghosn & Moses, 1986 and Nowak, 1999). Table 12 also lists the characteristic values according to Eurocode 1 for a bridge with a 9m deck-width. The traffic load effects obtained from the real Slovenia WIM data are more than 20% lower than the ones recommended by the Eurocode of actions.

This result shows the usefulness of using the proposed approach along with WIM data for evaluating the effect of traffic

Table 12. Comparison of characteristic values of maximum bending moment at mid-span from WIM data and Eurocode.

Span (m)	1000-year return value (kNm)	Eurocode LM1 (kNm)	Ratio
20	6,414	8,100	.79
30	10,930	13,725	.80
40	15,875	20,400	.78
50	20,374	28,125	.72

loads on existing bridges. The approach helps engineers estimate the actual safety of the bridge by modelling the actual traffic loads rather than using the code specified values which may condemn a bridge that otherwise is actually safe. It should be noted that this result is somewhat expected, as it is well known that truck traffic in Slovenia is not heavy compared to other European traffic. The emphasis here is placed on the benefits of using the proposed simple approach for particular cases rather than using the generic code-specified load model or using a complex simulation which would require an extensive programming and computing effort.

Conclusions

This paper reviewed the approach proposed by Ghosn and Sivakumar (2010) and NCHRP 12–76 (Sivakumar et al., 2011) for estimating the maximum truck load effect on bridges based on WIM data. The paper demonstrates that the simple approach yields results very similar to those of more sophisticated models with a fraction of the required computational time and with much simpler input data requirements.

A parametric analysis shows that the approach, which is based on fitting the tail end of WIM data results with an equivalent normal distribution, is not very sensitive to errors in well-calibrated WIM equipment. The method gives robust accurate results even when the WIM data are collected over short periods of time as long as the data collection period is carefully chosen to coincide with the crossing of representative truck traffic in terms of truck weight and volume.

The use of the proposed simple method, applicable to short- and medium-span bridges with span lengths less than 50 m, will help obtain improved assessments of the safety of existing bridges by providing realistic estimates of the applied live load effects when compared to the generic live load models recommended in codes and specifications oriented to bridge design.

Acknowledgements

The authors would like to thank the Spanish Ministries of Economy and Education for partially funding the study through Research Projects REHABCAR (INNPACTO) and BIA2010-16332 and Project SAB2009-0164 for supporting Prof. Ghosn's sabbatical leave at UPC. Thanks are also given to Mr Ales Znidaric from ZAG for providing the Slovenia WIM data

Table 11. Maximum moment in kNm for different service lives and span lengths.

	20 m			30 m			40 m			50 m		
	80 km/h	60 km/h	40 km/h	80 km/h	60 km/h	40 km/h	80 km/h	60 km/h	40 km/h	80 km/h	60 km/h	40 km/h
1 year	3656.7	3680.8	3806.7	6291	6472	6571.2	9318.1	9433.2	9625.8	12025.6	12161.8	12378.6
5 years	3854	3876.6	3995.2	6590.7	6761.2	6854.9	9711.6	9820.3	10002.5	12515.8	12644.6	12849.8
50 years	4116.8	4137.7	4247.5	6990.3	7148.3	7235.3	10238.3	10339.3	10509	13172.8	13292.7	13484.1
75 years	4161	4181.7	4290.1	7075.6	7213.6	7299.6	10327.1	10426.9	10594.6	13283.8	13402.2	13591.5

and his help in interpreting the traffic data and Mr Bala Sivakumar from HNTB for providing and helping analyse the U.S.A. WIM data. The financial support provided by NCHRP which sponsored parts of the work presented in this manuscript is also acknowledged.

Disclosure statement

No potential conflict of interest was reported by the authors.

Funding

This work was partially supported by the Spanish Ministries of Economy and Education through Research Projects REHABCAR (INNPACTO) and BIA2010 [16332] and Project [SAB2009-0164].

ORCID

Joan R. Casas ⓘ http://orcid.org/0000-0003-4473-4308

References

AASHTO. (2014). *LRFD bridge design specifications* (7th ed.). Washington, DC: American Association of State Highway and Transportation Officials.

AASHTO. (2011). *MBE manual for bridge evaluation* (2nd ed.). Washington, DC: American Association of State Highway and Transportation Officials.

Ang, A. H., & Tang, W. H. (2006). *Probability concepts in engineering planning and design*. New York, NY: Wiley.

ARCHES D-08. 2009. *Recommendations on the use of results of monitoring on bridge safety assessment and maintenance*. Deliverable D-08, ARCHES Project. VI EU Framework Program, Brussels. Retrieved from http://arches.fehrl.org.

Caprani, C., Grave, S. A., O'Brien, E., & O'Connor, A. 2002. Critical loading events for the assessment of medium span bridges. In Topping, B. H. V. & Bittnar, Z., (Eds.), ICCST '02: Proceedings of the 6th International Conference on Computational Structures Technology, Prague, Czech Republic: Civil Comp Press.

Casas, J. R. (1999). Evaluation of existing concrete bridges in Spain. *ACI Concrete International, 21*, 48–53.

Casas, J. R. (2000). Safety of prestressed concrete bridges to fatigue: Application to serviceability limit state of decompression. *ACI Structural Journal, 97*, 68–74.

Casas, J., Ghosn, M., & Soriano, M. (2012). *REHABCAR. Entregable 6.2.3. Guía para la definición de una prueba de carga tipo "proof". Software, ábacos y tablas para la obtención de la prueba de carga* [REHABCAR. Deliverable 6.2.3. Guideline to define a proof load test. Software, figures and tables to obtain the target proof load]. Madrid: Ministerio de Fomento, REHABCAR Project, INPACTO.

Casas, J. R., & Gomez, J. D. (2013). Load rating of highway bridges by proof-loading. *Journal of Civil Engineering, Korean Society of Civil Engineers, 17*, 556–567.

Cohen, H., Fu, G., Dekelbab, W., & Moses, F. (2003). Predicting truck load spectra under weight limit changes and its application to steel bridge fatigue assessment. *Journal of Bridge Engineering, 8*, 312–322.

Crespo-Minguillón, C., & Casas, J. R. (1997). A comprehensive traffic load model for bridge safety checking. *Structural Safety, 19*, 339–359.

Crespo-Minguillón, C., & Casas, J. R. (1998). Fatigue reliability analysis of prestressed concrete bridges. *Journal of Structural Engineering, 124*, 1458–1466.

Enright, B., O'Brien, E., & Dempsey, T. (2010). *Extreme traffic loading in bridges*. Proceedings of the Fifth International Conference on Bridge Maintenance, Safety and Management IABMAS2010, Philadelphia, USA.

Enright, B., & O'Brien, E. (2013). Monte Carlo simulation of extreme traffic loading on short and medium span bridges. *Structure and Infrastructure Engineering, 9*, 1267–1282.

Fiorillo, G. (2015). *Reliability and risk analysis of bridge networks under the effect of highway traffic load* (Ph.D dissertation). Department of Civil Engineering, The City College of New York, New York, NY, USA.

Fu, G., & Tang, J. (1995). Risk-based proof-load requirements for bridge evaluation. *Journal of Structural Engineering, 121*, 542–556.

Fu, G., & You, J. (2011). Extrapolation for future maximum load statistics. *Journal of Bridge Engineering, 16*, 527–535.

Frangopol, D., Goble, G., & Tan, N. (1992). *Truck loading data for a probabilistic bridge live load model*. Proceedings of the Sixth ASCE Specialty Conference. Denver, Colorado, USA.

Ghosn, M., & Moses, F. (1986). Reliability calibration of bridge design code. *Journal of Structural Engineering, 112*, 745–763.

Ghosn, M., Moses, F., & Gobieski, J. (1986). *Evaluation of steel bridges using in-service testing*. Transportation Research Record, TRR 1072. Washington, DC: TRB.

Ghosn, M., & Frangopol, D. (1996). *Site-specific live load models for bridge evaluation. Probabilistic mechanics and structural and geotechnical reliability*. Proceedings of the 7th ASCE Specialty Conference, Worcester, MA, USA.

Ghosn, M., & Sivakumar, B. (2010). *Using weigh-i-motion data for modeling maximum live load effects on highway bridges*. Proceedings of the Fifth International Conference on Bridge Maintenance, Safety and Management IABMAS2010, Philadelphia, PA, USA.

Ghosn, M., Sivakumar, B., & Miao, F. (2013). Development of state-specific load and resistance factor rating method. *Journal of Bridge Engineering, 18*, 351–361.

Gindy, M., & Nassif, H. (2006). *Multiple presence statistics for bridge live load based on weigh-in-motion data*. Transportation Research Board, 86th Annual Meeting, Washington, DC.

Grundy, P., & Boully, G. (2004). *Fatigue design in the new Australian bridge design code*. Proceedings of the Austroads 5th Bridge Conference, Hobart, Tasmania.

Jacob, B. 1991. *Methods for the prediction of extreme vehicular loads and load effects on bridges*. Report of subgroup 8. Eurocode 1: Traffic loads on bridges. Paris: LCPC.

Laman, J. A., & Nowak, A. S. (1996). Fatigue-load models for girder bridges. *Journal of Structural Engineering, 122*, 726–733.

Laman, J. A., & Nowak, A. S. (1997). Site-specific truck loads on bridges and roads. *Proceedings of the Institution of Civil Engineers: Transport, 123*, 119–133.

Lu, Q., Harvey, J., Le, T., Lea, J., Quinley, R., Redo, D., & Avis, J. (2002). *Truck traffic analysis using weigh-in-motion (WIM) data in California*. Berkeley, CA: Institute of Transportation Studies, Pavement Research Center, University of California.

Moses, F., Ghosn, M., & Snyder, R. E. (1984). Application of load spectra to bridge rating. *Transportation Research Record, 950*, 45–53.

Moses, F., & Ghosn, M. (1985). *A comprehensive study of bridge loads and reliability: Final report*. Columbus, OH: Federal Highway Administration and Ohio Department of Transportation.

Moses, F., Schilling, C. G., & Raju, K. S. (1987). *NCHRP Report 299: Fatigue evaluation procedures for steel bridges*. Washington, DC: Transportation Research Board.

Moses, F. (2001). *NCHRP Report 454, Calibration of Load Factors for LRFR Bridge Evaluation*. Washington, DC: Transportation Research Board.

Nowak, A. S., & Nassif, H. (1992). *Live load models based on WIM data. Probabilistic Mechanics and Structural and Geotechnical Reliability*. Proceedings of the 6th ASCE Specialty Conference, Denver, CO, USA.

Nowak, A. S. (1999). *NCHRP Report 368, Calibration of LRFD Bridge Design Code*. Washington, DC: Transportation Research Board.

O'Brien, E., & Znidaric, A. (2001). *Report of Work Package 1.2 – Bridge WIM Systems (B-WIM)*. European project WAVE. Brussels: European Commission.

O'Brien, E., Schmidt, F., Hajializadeh, D., Zhou, X., Enright, B., Caprani, C., … Sheils, E. (2015). A review of probabilistic methods of assessment of load effects in bridges. *Structural Safety, 53*, 44–56.

O'Connor, A., Jacob, B., O'Brien, E., & Prat, M. (1998). *Effects of traffic loads on road bridges – Preliminary studies for the re-asessment of the traffic load model for Eurocode 1, Part 3*. 2nd European Conference on Weigh-in-Motion of Road Vehicles, Lisbon. European Commission, Luxembourg.

Sivakumar, B., Ghosn, M., & Moses, F. (2011). *Protocols for collecting and using traffic data in bridge design*. NCHRP Report 683. Washington, DC: Transportation Research Board, The National Academies.

Treacy, M., Brühwiler, E., & Caprani, C. (2014). Monitoring of traffic action local effects in highway bridge deck slabs and the influence of measurement duration on extreme value estimates. *Structure and Infrastructure Engineering, 10*, 1555–1572.

Wang, T.-L., Liu, C., Huang, D., & Shahawy, M. (2005). Truck loading and fatigue damage analysis for girder bridges based on weigh-in-motion data. *Journal of Bridge Engineering, 10*, 12–20.

Wiśniewski, D., Casas, J. R., & Ghosn, M. (2012). Codes for safety assessment of existing bridges – Current state and further development. *Structural Engineering International, 22*, 552–561.

Znidaric, A., Lavric, I., Kalin, & J. (2010). *Latest practical developments in the Bridge WIM Technology*. Proceedings of the Fifth International Conference on Bridge Maintenance, Safety and Management IABMAS2010. Philadelphia, USA.

Znidaric, A., Kreslin, M., Lavric, I., & Kalin, J. (2012). Simplified approach to modeling traffic loads on bridges. Transport Research Arena. *Social and Behavioral Sciences, 48*, 2887–2896.

Reliability-based durability design and service life assessment of reinforced concrete deck slab of jetty structures

Mitsuyoshi Akiyama, Dan M. Frangopol and Koshin Takenaka

ABSTRACT

Reinforced concrete (RC) structures in a marine environment deteriorate with time due to chloride-induced corrosion of reinforcing bars. Since the RC deck slabs of jetty structures are exposed to a very aggressive environment, higher deterioration rates can develop. In this paper, a reliability-based durability design and service life assessment of RC jetty structures are presented. For new RC jetty structures, the concrete quality and concrete cover necessary to prevent the chloride-induced reinforcement corrosion causing the deterioration of structural performance during the whole lifetime could be determined. Based on the airborne chloride hazard depending on the vertical distance from the sea level surface to the RC deck slab, the probability associated with the steel corrosion initiation is estimated. The water to cement ratio and concrete cover to satisfy the target reliability level are provided. For evaluating the service life of existing structures, the condition state based on the visual inspection of RC structure can be provided. The deterioration process of the RC jetty structure can be modelled as a Markov process. Therefore, the transition probability matrix at time t after construction can be updated by visual inspection results. A procedure to update the transition probability matrix by the Sequential Monte Carlo Simulation method is indicated. In an illustrative example, the effect of the updating on the life-cycle reliability estimate of existing RC deck slab in a jetty structure subjected to the chloride attack is presented.

1. Introduction

For structures located in a moderately or highly aggressive environment, multiple environmental and mechanical stressors lead to deterioration of structural performance. Despite the fact that concrete is a reliable structural material with good durability performance, exposure to severe environments makes it vulnerable (Moradi-Marani, Shekarchi, Dousti, & Mobasher, 2010). Reinforcement corrosion in concrete is the dominant cause of premature deterioration of reinforced concrete (RC) structures. The corrosion-induced structural deterioration is a gradual process consisting of a few phases during the service life of RC structures, including corrosion initiation, cover concrete cracking, serviceability loss, and collapse due to loss of structural strength. Such deterioration will reduce the service life of structures and increase their life-cycle cost due to maintenance actions. Whole-life performance prediction of RC structures is gradually becoming a requirement for the design of these structures and a necessity for optimal decision-making with respect to their inspection, repair, strengthening, replacement and demolition.

Despite extensive research in this field, a number of issues still remain unclear. One of the main intricacies is the uncertainty associated with the physical parameters involved in the problem.

Because of the presence of many kinds of uncertainties associated with the prediction of deterioration process of RC structures, various theoretical frameworks have been developed to assess the service life performance of structures, including advanced reliability theories (Ellingwood, 2005; Frangopol, 2011; Li, 2003, 2004; Mori & Ellingwood, 1993).

In this paper, the reliability-based durability design and service life assessment of RC deck slab of jetty structures are presented. Figure 1 shows a typical RC jetty in Tokyo port of Japan. Dousti, Moradian, Taheri, Rashetnia and Shekarchi (2013) reported, based on their field investigation, that the assessment of RC jetty structures showed that, when the structures are exposed to aggressive environments, very high deterioration rates can develop, leading to serious damage conditions in very short time periods. The highest deterioration rates were observed in zones where the concrete surface is subjected to perpetual wetting and drying cycles of salt water.

For new RC deck slab in a jetty structure, it is possible to determine the water to cement ratio and concrete cover to prevent the steel corrosion initiation during the whole lifetime of the structure under various uncertainties by conducting an adequate time-dependent reliability analysis. Several marine environmental factors affect the degradation mechanisms of concrete structures. The exact influence of these factors is difficult to predict

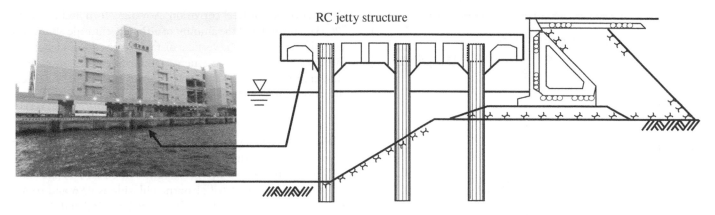

RC jetty structure

Figure 1. RC jetty structure in Tokyo port of Japan (Source: Author).

as they vary in time and space. Unlike the case associated with actions that are usually considered in structural design (e.g. probabilistic seismic hazard analysis and traffic load simulation), there is a lack of research on marine environmental hazard assessment. This is because the data on coastal atmospheric exposure is very limited compared with the array measurements of seismic ground motion. In the seismic probabilistic hazard analysis (SPHA), which determine the relation between an earthquake motion intensity (e.g. PGA, PGV and response acceleration) and the corresponding annual probability of exceedance, it is necessary to evaluate the attenuation of seismic intensity with the distance from the seismic source (Cornell, 1968). By applying the concept of SPHA to the assessment of the effect of airborne chlorides on durability design of RC structures, which is reduced with increasing the distance from the coastline, probabilistic hazard analysis can be performed and the effect of a marine environment can be quantified. Akiyama et al. (Akiyama, Frangopol, & Suzuki, 2012; Akiyama, Frangopol, & Matsuzaki, 2014) proposed an approach to establish the probabilistic hazard curve associated with airborne chlorides depending on the horizontal distance from the coastline and provided a computational procedure to obtain the probability of occurrence of steel corrosion and corrosion cracking. Since RC jetty structures are located above the sea, it is necessary to evaluate the hazard associated with airborne chlorides depending on the vertical distance from the sea-level surface. The procedure to estimate the failure probability associated with the steel corrosion initiation for RC deck slab in the jetty structure is presented in this paper.

Because of the lack of adequate knowledge to ensure the durability of concrete structures, existing RC jetty structures have significantly deteriorated. Investigation of corrosion damage in RC jetty structures is mainly focused on the RC deck slab. A systematic visual inspection is conducted to get a realistic data about the deterioration and distress of the concrete deck slabs and to take the extensive photographs of the deck slabs for determining the extent of investigation (Dousti et al. 2013). Based on these visual inspections, the probabilities of transition between condition states of RC deck slab are provided. These probabilities could be incorporated into the Markov chain.

Markov chains are commonly used for performance assessment of deteriorating components and systems. The rate of transition from one state to another is constant over time for homogeneous Markov chains. Madanat (1993) presented a

methodology for planning the maintenance and rehabilitation activities of transportation facilities based on the latent Markov decision process. Kato, Iwanami, Yokota and Yamaji (2002) applied the Markov model to the life-cycle reliability estimate of the RC jetty structures. Recently, Saydam, Frangopol and Dong (2013) presented a methodology for quantifying lifetime risk of bridge superstructures. In their model, a scenario-based approach integrating the Pontis element condition rating system into risk assessment procedure was used to identify expected losses, and a Markov process was applied to estimate the deterioration level of bridge components regarding the transition between the condition states. Reliability estimation of RC structures under hazard associated with environmental stressors and material deterioration models requires complex computations including hazard and fragility analyses (e.g. Akiyama, Frangopol, & Yoshida, 2010). Since in Markov model, deterioration process can be expressed as the transition probability, it can provide significant time efficiency in life-cycle reliability analysis.

The transition probability matrix used in Markov model depends on the evaluation of environment, corrosion process of steel bars and deterioration of structural performance. The model uncertainties associated with the estimation of the transition probability matrix could be very large if there is only information on design condition of existing RC structures analysed. Meanwhile, for existing structures it is possible to reduce epistemic uncertainties using inspection results (Akiyama et al. 2010; Yoshida, 2009). In this study, a computational procedure for estimating the life-cycle reliability of existing RC jetty structures subjected to the chloride attack is presented. The Markov model and Sequential Monte Carlo Simulation (SMCS) are both used in conjunction with time-dependent condition state probability for existing RC jetty structures. An illustrative example is presented.

2. Reliability-based durability design of RC deck slab of a jetty structure

2.1. Hazard assessment associated with airborne chloride

Environmental conditions should be quantitatively assessed and the evaluation results should be reflected in the rational durability design of RC structures in a marine environment. The difference in the amount of airborne chlorides among structure locations

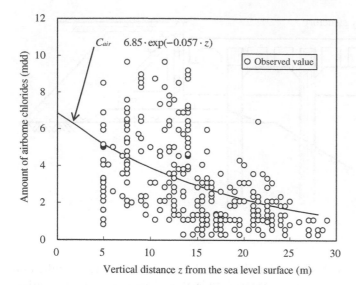

Figure 2. Amount of airborne chlorides vs. vertical distance from the sea level surface (based on the observed values reported by Aoyama, Torii and Matsuda, 2003).

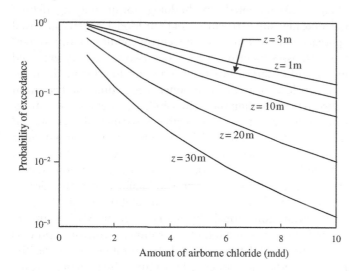

Figure 3. Probability profiles for amounts of airborne chlorides depending on the vertical distance z from the sea level surface.

should be considered taking into account the spatial-temporal variation.

Akiyama et al. (2012, 2014) proposed the attenuation relation between the amount of observed airborne chlorides, C_{air}, and the horizontal distance from the coastline. The speed of wind, the ratio of sea wind (defined as the percentage of time during one day when the wind is blowing from sea toward land) and the distance from the coastline affect the amount of airborne chlorides. C_{air} (mdd = 100 mg/m²/day) in the lateral direction can be expressed as:

$$C_{air} = 1.29 \cdot r \cdot u^{0.386} \cdot d^{-0.952} \quad (1)$$

where r is the ratio of sea wind, u (m/s) is the average wind speed during the observation period and d (m) is the distance from the coastline. The statistics of wind speed and the ratio of sea wind of C_{air} collection could be obtained from meteorological data.

Since RC jetty structures are located above the sea, Equation (1) cannot be applied to estimate the probability associated with

the occurrence of steel corrosion. Aoyama, Torii and Matsuda (2003) reported that the amount of airborne chloride attenuates with the increase in the vertical distance from the sea level surface based on the measurements of chloride concentration as shown in Figure 2. The attenuation of C_{air} (mdd) in the vertical direction can be expressed as:

$$C_{air} = 6.85 \cdot \exp\left(-0.057 \cdot z\right) \quad (2)$$

where z (m) is the vertical distance of the bottom of RC deck slab from the sea level surface.

The mean and coefficient of variation (COV) of the ratio of observed to calculated airborne chloride is 0.96 and 0.646, respectively. Since the effect of the speed of wind and the ratio of sea wind on the amount of airborne chloride is negligible, the attenuation of C_{air} in Equation (2) depends only on the distance from the sea level surface. Because of large uncertainty involved in the prediction of airborne chloride spread in the vertical direction, larger concrete cover and smaller water to cement ratio are needed to ensure the prescribed target reliability level. In order to reduce this uncertainty, further research is needed to include additional parameters in the attenuation equation.

The probability that C_{air} at a specific site will exceed a prescribed value c_{air} is:

$$
\begin{aligned}
q_s\left(c_{air}\right) &= P\left(C_{air} > c_{air}\right) \\
&= P\left(U_R > \frac{c_{air}}{6.85 \cdot \exp\left(-0.057z\right)}\right)
\end{aligned}
\quad (3)
$$

where U_R is the lognormal random variable representing attenuation uncertainty associated with Equation (2).

Figure 3 shows the probability profiles (i.e. hazard curve) associated with the airborne chloride. Unlike the airborne chloride probability profiles depending on the horizontal distance from the coastline (Akiyama et al. 2012, 2014), those shown in Figure 3 are independent of the location of the structure.

2.2. Reliability assessment of RC deck slab in a jetty structure

When airborne chlorides were measured in each location, exposed concrete specimens were also placed at the same location to measure the surface chloride content, C_0. Based on the measurements of chloride concentration (Aoyama 2003), the relationship between C_{air} (mdd) and C_0 (kg/m³) is:

$$C_{air} = 0.563 \cdot C_0^{0.948} \quad (4)$$

The fragility curve shows the occurrence probability of the limit state under the condition that a specific value of airborne chloride, c_{air} is given. The probability of occurrence of steel corrosion given c_{air} can be obtained using the performance function (Akiyama et al. 2012):

$$g_{d,1} = C_T - \chi_1 \cdot C(c, \chi_2 D_c, \chi_3 C_0, t) \quad (5)$$

where $\quad \log D_c = -6.77(W/C)^2 + 10.10(W/C) - 3.14 \quad (6)$

$$C(c, \chi_2 D_c, \chi_3 C_0, t) = \chi_3 C_0 \left\{ 1 - \mathrm{erf}\left(\frac{0.1 \cdot c}{2\sqrt{\chi_2 D_c t}}\right) \right\} \quad (7)$$

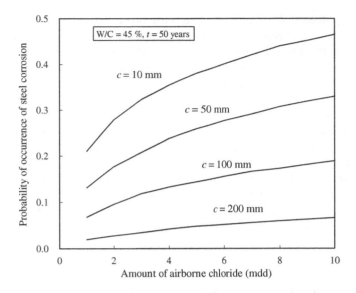

Figure 4. Effect of concrete cover on the probability profiles associated with the steel corrosion assuming W/C = 50% and t = 50 years after construction.

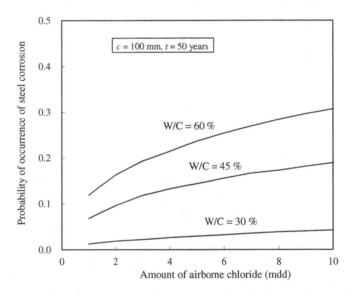

Figure 5. Effect of water to cement ratio on the probability profiles associated with the steel corrosion assuming c = 100 mm and t = 50 years after construction.

where C_T is the critical threshold of chloride concentration (kg/m^3), c is the concrete cover (mm), t is the time after construction (years), W/C is the ratio of water to cement, erf is the error function, χ_1 is the model uncertainty associated with the estimation of C, D_c is the coefficient of diffusion of chloride, χ_2 is the model uncertainty associated with the estimation of D_c and χ_3 is the model uncertainty associated with Equation (6).

The parameters of random variables were provided in Akiyama et al. (2012). Figures 4 and 5 show the fragility curves $F_r(c_{air})$. The effects of concrete cover and water to cement ratio on the probabilities of occurrence of steel corrosion given c_{air} are presented in Figures 4 and 5, respectively. Using the fragility curve, $F_r(c_{air})$, the probability of steel corrosion initiation is estimated by:

$$p_f = \int_0^{\infty} \left(-\frac{dq_s(c_{air})}{dc_{air}} \right) \cdot F_r(c_{air}) \, dc_{air} \quad (8)$$

Figures 6 and 7 illustrate the relationship between the probability of steel corrosion initiation and distance z from the sea level surface assuming RC deck slab with W/C = 45% and t = 10 years, and W/C = 45% and t = 50 years, respectively. Using Equation (8), the concrete cover and water to cement ratio to ensure the target reliability could be determined; however, this would require the structural designer to conduct the reliability computations.

2.3. Reliability-based durability design

A design criterion to ensure that the reliability index for the occurrence of steel corrosion will be close to the target value, without performing a complex reliability analysis by the designers, is proposed herein. The proposed formulation is:

$$\gamma \frac{C_d}{C_{T,d}} \leq 1.0 \quad (9)$$

$$C_d = C_{0,d} \left(1 - \text{erf} \frac{0.1 \cdot c}{2 \sqrt{D_{c,d} \cdot T}} \right) \quad (10)$$

$$C_{0,d} = 4.5 \cdot \exp(-0.05 \cdot z) \quad (11)$$

where γ is the durability design factor, $C_{T,d}$ is the design critical threshold of chloride concentration (C_T in Equation (5) as the mean value), and T is the lifetime of RC jetty structure.

The procedure to determine the durability design factor is based on code calibration. The steps are as follows (Akiyama et al. 2012, 2014):

(a) Set the target reliability index β_{target} and the lifetime of the structure T.
(b) Calculate the design value of the surface chloride content using Equation (11).
(c) Assume the initial durability design factor γ and the distance z from the sea level surface.
(d) Select the design concrete cover and water to cement ratio to satisfy the design criterion (i.e. Equation (9)).
(e) Calculate the probability using Equation (8) and transform it into reliability index.
(f) Repeat steps (c) to (e) until:

$$U = \sum_{i=1} \left(\beta_{target} - \beta_i(\gamma) \right)^2 \quad (12)$$

is minimised, and the durability design factor is found.

When T = 50 years, the durability design factors associated with the target reliability indices 1.0, 1.5 and 2.0 are 1.07, 3.45 and 12.9, respectively. Figure 8 indicates the reliability index associated with the steel corrosion initiation assuming that the concrete with W/C = 30% is used for the RC jetty structure. The results indicate that the reliability indices are close to the target values. Using the design criterion and durability design factor proposed, RC jetty structures having target durability reliability indices for prescribed lifetime T can be designed.

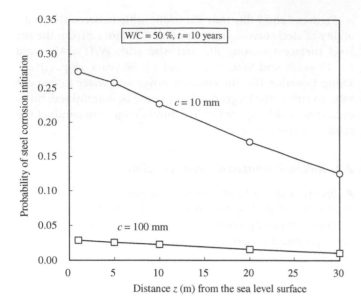

Figure 6. Relationship between probability of steel corrosion initiation and distance z from the sea level surface assuming W/C = 50% and t = 10 years after construction for two concrete covers, c = 10 mm and c = 100 mm.

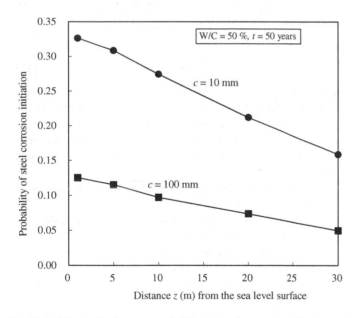

Figure 7. Relationship between probability of steel corrosion initiation and distance z from the sea level surface assuming W/C = 50% and t = 50 years after construction for two concrete covers, c = 10 mm and c = 100 mm.

3. Life-cycle reliability estimation of existing RC deck slab of a jetty structure

RC deck slab of a jetty structure with high quality concrete and adequate concrete cover designed by Equation (9) prevents the chloride-induced reinforcement corrosion causing the deterioration of structural performance during its whole lifetime. Based on the life-cycle cost analysis, corrosion-resistant stainless steel reinforcing bars would be used to protect RC jetty structure from the chloride attack, even if it is much more expensive than conventional carbon steel (Val & Stewart, 2003). Although cost estimation is outside of the scope of this paper, life-cycle reliability analysis can be applied to select optimal strategies for improving durability of new RC jetty structures.

Meanwhile, some existing structures designed without adequate durability detailing deteriorate severely. To confirm whether these deteriorated structures still conform to the safety and/or serviceability requirements, it is necessary to investigate the effect of the chloride-induced reinforcement corrosion on the structural capacity and stiffness. Although the life-cycle reliability assessment of existing corroded RC structures has been developed, it requires complex reliability computations. For existing RC jetty structures, it is important to identify when the maintenance actions including the repair and/or strengthening are needed. Markov model is an efficient tool to estimate the time-variant performance of structures. The accuracy of Markov model in the life-cycle reliability analysis depends strongly on the transition probability which is the probability of moving from any given state to the subsequent state on the next time interval. The uncertainty associated with the estimation of the transition probability becomes large if the transition probability is determined based on the information on the design condition of RC jetty structures. Using inspection results, the epistemic uncertainties associated with the service life reliability prediction of existing RC structures can be reduced compared with new RC structures.

In this section, the deterioration process of a RC jetty structure transitioning between condition states is modelled as a Markov process. A procedure to obtain the transition probability matrix at time t after construction updated by SMCS is indicated. Transition probability matrix could be updated in order to be consistent with the observational information.

3.1. Markov model

The rate of transition from one state to another is assumed to be constant over time for homogeneous Markov chains. The probability of moving from any given state $(k-1)$ to state k on the next time interval is called the transition probability p_k. This probability is the core of a Markov chain. In this paper, a RC jetty structure in a marine environment with six condition states (i.e. k = 0, 1, 2, 3, 4 and 5) is considered. The deterioration states of RC jetty structures due to chloride-induced corrosion of reinforcing bars are classified according to the criteria defined by the Japan Society of Civil Engineers (JSCE, 2001) and Komure, Hamada, Yokota, and Yamaji (2002) as listed in Table 1. The relationship between the state probability vector $X(t)$ and the transition probability matrix that can be used in the prediction of the structural performance is:

$$
\begin{Bmatrix} X_0 \\ X_1 \\ X_2 \\ X_3 \\ X_4 \\ X_5 \end{Bmatrix} = \begin{Bmatrix} 1-p_1 & 0 & 0 & 0 & 0 & 0 \\ p_1 & 1-p_2 & 0 & 0 & 0 & 0 \\ 0 & p_2 & 1-p_3 & 0 & 0 & 0 \\ 0 & 0 & p_3 & 1-p_4 & 0 & 0 \\ 0 & 0 & 0 & p_4 & 1-p_5 & 0 \\ 0 & 0 & 0 & 0 & p_5 & 1 \end{Bmatrix}^t \begin{Bmatrix} 1 \\ 0 \\ 0 \\ 0 \\ 0 \\ 0 \end{Bmatrix}
$$

$$
= \begin{Bmatrix} f_0(p_1,t) \\ f_1(p_1,p_2,t) \\ f_2(p_1,p_2,p_3,t) \\ f_3(p_1,p_2,p_3,p_4,t) \\ f_4(p_1,p_2,p_3,p_4,p_5,t) \\ f_5(p_1,p_2,p_3,p_4,p_5,t) \end{Bmatrix}
\tag{13}
$$

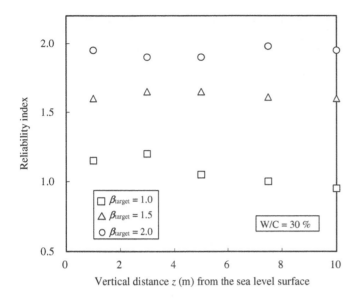

Figure 8. Reliability index of RC deck slab in a jetty structure designed by Equation (10) assuming W/C = 30%.

Table 1. Deterioration states of RC jetty defined by Japan Society of Civil Engineers (JSCE, 2001) and Komure, Hamada, Yokota and Yamaji (2002).

Condition state	Deterioration status
0	Sound
1	Some rusts are observed on the concrete surface. Crack width is less than 0.3 mm
2	Minor corrosion cracks are shown. Crack width is larger than 0.3 mm
3	Major corrosion cracks are shown. Crack width is larger than 1.0 mm
4	Deteriorated seriously. More than 10% of cover concrete of RC deck slab are spalled. Structural performance declined
5	Deteriorated totally. More than 40% of cover concrete of RC deck slab are spalled. Obvious decline of structural performance

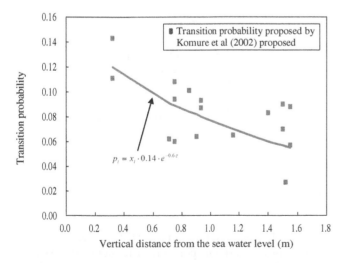

Figure 9. Relationship between transition probability and distance from the sea water level to the RC deck slab in the jetty structure.

where $X(t) = [X_0, X_1, X_2, X_3, X_4, X_5]$.

Komure et al. (2002) proposed the annual transition probability to quantify the actual chloride-induced damage state based on the survey results of many existing RC jetty structures around Japan. To prevent moving from any given state $(k-1)$ to state $k+1$, the time interval in Equation (13) was assumed to be one year. Figure 9 shows the relationship between transition probability and the vertical distance from the seawater level to the RC deck slab. Assuming that transition probability is independent of conditional states (i.e. $p_1 = p_2 = \ldots = p_5$), p_i is provided by:

$$p_1 = p_2 = p_3 = p_4 = p_5 = 0.14 \cdot e^{-0.6 \cdot z} \tag{14}$$

When model uncertainty associated with the estimation of transition probability p_i is taken into consideration, p_i is provided by

$$p_i = x_i \cdot 0.14 \cdot e^{-0.6 \cdot z} \tag{15}$$

where x_i is the model uncertainty, assumed to follow a lognormal distribution, associated with the estimation of p_i.

3.2. Sequential Monte Carlo Simulation

For existing structures, the uncertainties associated with predictions can be reduced by the effective use of information obtained from visual inspections, field test data regarding structural performance, and/or monitoring (Enright & Frangopol, 1999; Estes & Frangopol, 2003; Estes, Frangopol, & Foltz, 2004; Frangopol, 2011; Frangopol & Liu, 2007; Frangopol, Saydam, & Kim, 2012; Frangopol, Strauss, & Kim, 2008a, 2008b, Okasha & Frangopol, 2009; Okasha, Frangopol, & Orcesi, 2012). This information helps engineers to improve accuracy of long-term structural performance estimation. In this paper it is assumed that the state probability vector at t years is given based on the survey of existing RC jetty structures. Since the relationship between the state probability vector and related random variables is nonlinear as described in Equations (13–15), it is impossible to perform the updating of these random variables by a theoretical closed-form solution such as a Kalman Filter algorithm.

In this paper, SMCS is applied to update the random variables x_i. The state space model consists of two processes, the time updating process and the observation updating process. The time

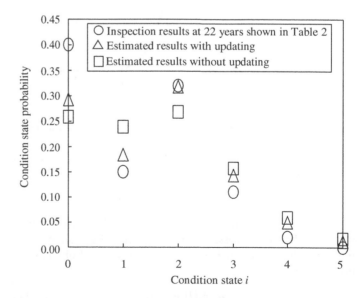

Figure 10. Comparison of inspection results at 22 years, with estimated results with updating at 16 years and estimated results without updating at 16 years.

Table 2. Probabilities of condition state based on inspection of an existing RC jetty structure (adapted from Taniguchi, Tamura, Sano, and Hamada 2004).

Condition state	Probability of condition state					
	0	1	2	3	4	5
16 years after construction	0.42	0.18	0.30	0.08	0.02	0.00
22 years after construction	0.40	0.15	0.32	0.11	0.02	0.00

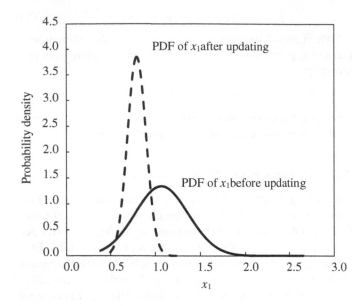

Figure 11. Comparison of the PDFs of x_1 with updating at 16 years and without updating.

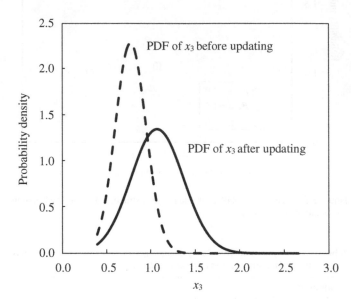

Figure 13. Comparison of the PDFs of x_3 with updating at 16 years and without updating.

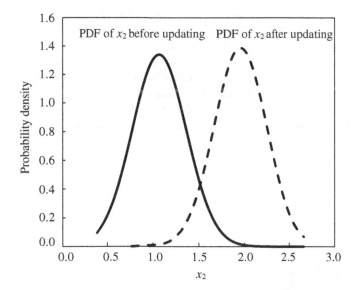

Figure 12. Comparison of the PDFs of x_2 with updating at 16 years and without updating.

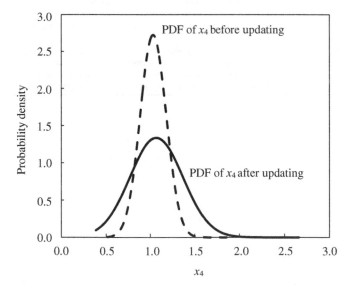

Figure 14. Comparison of the PDFs of x_4 with updating at 16 years and without updating.

updating process is the one step ahead prediction based on the information at the $(k-1)$-th step. The predicted vector is:

$$x_{k/k-1} = F(x_{k-1/k-1}, w_k) \qquad (16)$$

where w_k is the system noise represented by the noise involved in the prediction process. It is assumed that observation information z_k is a function H of state vector $x_{k/k}$ and observation noise v_k as:

$$z_k = H(x_{k/k}, v_k) \qquad (17)$$

The probability density functions (PDFs) of these noises are assumed known and independent. The algorithm based on MCS starts by assuming samples drawn from the distribution at $(k-1)$-th step:

$$x_{k-1/k-1}^{(j)} \sim p(x_{k-1}|Z_{k-1}), \quad j = 1, \dots, n \qquad (18)$$

$$Z_{k-1} = (z_1, z_2, \dots, z_{k-1}) \qquad (19)$$

The superscript (j) denotes the generated j-th sample realisation. The PDF is approximately expressed by the samples with Dirac delta function δ as:

$$p(x_{k-1}|Z_{k-1}) \cong \frac{1}{n}\sum_{j=1}^{n}\delta\left(x_{k-1}-x_{k-1/k-1}^{(j)}\right) \qquad (20)$$

The above approximation form of PDF is called empirical PDF. The samples of the k-th step before observation updating are obtained by simply substituting them into Equation (16) to become:

$$x_{k/k-1}^{(j)} = F\left(x_{k-1/k-1}^{(j)}, w_k^{(j)}\right) \qquad (21)$$

The empirical PDF of the k-th step before updating is similarly estimated by the sample realisation:

$$p(x_k|Z_{k-1}) \cong \frac{1}{n}\sum_{j=1}^{n}\delta\left(x_k-x_{k/k-1}^{(j)}\right) \qquad (22)$$

The PDF after updating is:

$$\begin{aligned}p(x_k|Z_k) &= p(x_k|z_k, Z_{k-1})\\ &= \frac{p(x_k, z_k|Z_{k-1})}{p(z_k|Z_{k-1})}\\ &= \frac{p(z_k|x_k, Z_{k-1})\cdot p(x_k|Z_{k-1})}{\int p(z_k|x_k, Z_{k-1})\cdot p(x_k|Z_{k-1})\cdot dx_k}\end{aligned} \qquad (23)$$

Substituting Equations (22) into (23) and using the property of a delta function results in:

$$\begin{aligned}p(x_k|Z_k) &= \sum_{j=1}^{n}\left[\frac{q_k^{(j)}}{\sum_{i=1}^{n}q_k^{(i)}}\right]\cdot\delta(x_k-x_{k/k-1}^{(j)})\\ &= \sum_{j=1}^{n}a_k^{(j)}\cdot\delta(x_k-x_{k/k-1}^{(j)})\end{aligned} \qquad (24)$$

where:

$$q_k^{(j)} = p(z_k|x_{k/k-1}^{(j)}) \qquad (25)$$

$$a_k^{(j)} = \frac{q_k^{(j)}}{\sum_{i=1}^{n}q_k^{(i)}} \qquad (26)$$

The term $a_k^{(j)}$ is the weight (likelihood ratio) of sample j. When a new observation is available, the weights are recalculated and the approximate posterior PDF is sequentially updated. The detailed procedure of SMCS applied to reliability analysis of concrete structure in a marine environment was given by Yoshida (2009).

3.3. Illustrative example

Based on the visual inspections, Taniguchi, Tamura, Sano and Hamada (2004) reported the difference of condition state probabilities of an existing RC deck slab in a jetty structure subjected to the chloride attack between 16 and 22 years after construction as listed in Table 2. RC jetty structures deteriorate due to the chloride attack, and, therefore, the probability of severe condition

states increases with time. Based on the inspection result at 16 years, the condition state at 22 years is predicted using the Markov model and SMCS.

Since computational results are almost the same if the number of samples N is more than 10,000, N in SMCS is set to 10,000. Based on the observation data, Equation (17) becomes:

$$\begin{pmatrix}Z_0\\Z_1\\Z_2\\Z_3\\Z_4\\Z_5\end{pmatrix} = \begin{pmatrix}X_0\\X_1\\X_2\\X_3\\X_4\\X_5\end{pmatrix} + \begin{pmatrix}v_0\\v_1\\v_2\\v_3\\v_4\\v_5\end{pmatrix}$$

$$= \begin{pmatrix}f_0(p_1, t)\\f_1(p_1, p_2, t)\\f_2(p_1, p_2, p_3, t)\\f_3(p_1, p_2, p_3, p_4, t)\\f_4(p_1, p_2, p_3, p_4, p_5, t)\\f_5(p_1, p_2, p_3, p_4, p_5, t)\end{pmatrix} + \begin{pmatrix}v_0\\v_1\\v_2\\v_3\\v_4\\v_5\end{pmatrix} \qquad (27)$$

where Z_i represents the the inspection results of condition state i, X_i is the prediction of each condition state i and v_i is the observation noises at condition state i.

Figure 10 compares the probabilities of the six condition states i ($i = 0, 1, 2, 3, 4\ 5$) reported in Table 2, with estimated results without and with updating at 16 years. The difference between inspection results and predictions from the proposed model

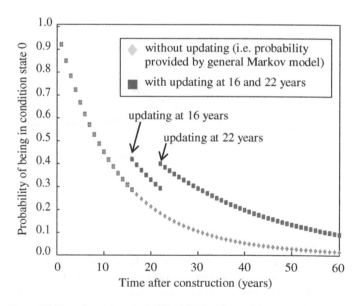

Figure 15. Time-dependent probability of being in condition state 0 with and without updating.

Table 3. List of assumed observation data.

X ($t = 20$ year)	Probability	
	Case 1	Case 2
X_0	0.44	0.19
X_1	0.37	0.33
X_2	0.14	0.27
X_3	0.04	0.14
X_4	0.01	0.05
X_5	0.00	0.02

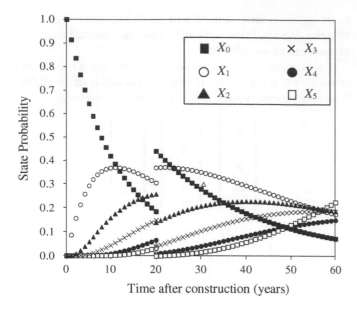

Figure 16. Time-dependent state probability (Case 1).

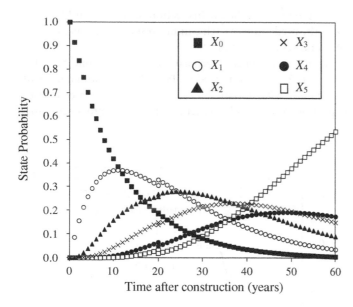

Figure 17. Time-dependent state probability (Case 2).

To examine the effect of observation information on the time-dependent condition state probability, two cases of state probability vector $X(t)$ determined by the visual inspection as listed in Table 3 are considered. It is assumed that the observation information is provided at 20 years after construction. Figures 16 and 17 show time-dependent condition state probabilities using Markov model and SMCS. As shown in Table 3, since the assumed probability of condition state 0 of RC jetty structure in Case 1 is larger than that in Case 2, X_5 at 60 years in Case 1 is smaller than that in Case 2. The condition state probability after updating depends on the additional information. It is updated in order to be consistent with inspection results.

At the design stage, life-cycle reliability assessment of new RC jetty structure is conducted using the transition probability p_k. This probability could be determined based on the survey results acquired from many structures over time, for example, Equation (14). The general the trend in the time-dependent reliability of RC jetty structures could be estimated. However, a different value of p_k needs to be used to represent the deterioration of different classes of structures under various environmental conditions. In the case of available adequate inspection data from the specific structure, p_k has to be updated for consistency with the inspection results.

4. Conclusions

The findings of the present study can be summarised as follows:

(1) Based on the attenuation of the amount of observed airborne chloride in the vertical direction, probability profiles associated with airborne chloride for the life-cycle reliability assessment of RC deck slab in the jetty structure were proposed. These profiles can quantify the effect of marine environment.

(2) A reliability-based method for durability design of RC jetty structures using a durability design factor was introduced. For new RC jetty structures, concrete quality and concrete cover to prevent the chloride-induced reinforcement corrosion during the whole lifetime of these structures could be determined using the proposed durability design method. RC structures could be designed so that the time taken for the occurrence of the steel corrosion provided by the partial factor and design criterion is longer than the design lifetime. Using this design method, RC jetty structures can satisfy the target reliability level without any reliability computations performed by structural designers.

(3) A life-cycle reliability estimation method using Markov chain and SMCS was presented for existing RC deck slab in a jetty structure. The structural deterioration is modelled using condition state probabilities (i.e. Markov chain state probabilities) and the transition probability matrix is updated by SMCS. It is important to have inspection results in order to reduce the epistemic uncertainties. This can help in performing a more accurate life-cycle reliability estimate of existing RC jetty structures.

(4) Further research is needed on the reliability of RC jetty structures in a marine environment in connection

using SMCS is smaller than that between inspection results and Equation (13) at $t = 22$ years. The proposed model using SMCS could perform a more accurate life-cycle reliability estimate of the RC jetty structure. Figures 11–14 show the effect of updating on the PDF of model uncertainty associated with the estimation of transition probability. The COV after updating is smaller than that without updating.

Figure 15 shows the time-dependent probability of condition state 0 with updating at 16 and 22 years. To examine the effect of updating on the probability of condition state 0, the time-dependent probability estimated by the general Markov model is also shown in Figure 15. At 60 years, the difference between the probabilities of condition state 0 with and without updating is not negligible. It is important to update the conditional probabilities using inspection results for establishing the rational maintenance strategy.

with probabilistic nonlinear FE modeling (Teigen et al. 1991a, 1991b), reliability-based inspection optimization (Onoufriou & Frangopol 2002), and life-cycle maintenance and optimisation (Frangopol & Kong 2001, Frangopol & Tsompankis 2014). Further research is necessary on life-cycle performance of RC jetty structures by improving durability, reducing maintenance costs and extending service life (e.g. use of high-performance concrete, admixtures and/or corrosion-resistant stainless steel rebars).

Acknowledgements

The opinions and conclusions presented in this paper are those of the authors and do not reflect the views of the sponsoring organisations.

Disclosure statement

No potential conflict of interest was reported by the authors.

Funding

This work was supported by JSPS Grant-in-Aid for Scientific Research (B) [grant number 24360185], and the Exploratory Research [grant number 25630198].

References

Akiyama, M., Frangopol, D. M., & Matsuzaki, H. (2014). Reliability-based durability design and service life assessment of concrete structures in an aggressive environment. Chapter 1. In D.M. Frangopol & Y. Tsompankis (Eds.), *Maintenance and safety of aging infrastructure* (pp. 1–26). London: CRC Press/Balkema, Taylor & Francis Group.

Akiyama, M., Frangopol, D. M., & Suzuki, M. (2012). Integration of the effects of airborne chlorides into reliability-based durability design of reinforced concrete structures in a marine environment. *Structure and Infrastructure Engineering, 8*, 125–134.

Akiyama, M., Frangopol, D. M., & Yoshida, I. (2010). Time-dependent reliability analysis of existing RC structures in a marine environment using hazard associated with airborne chlorides. *Engineering Structures, 32*, 3768–3779.

Aoyama, M., Torii, K., & Matsuda, T. (2003). A study on the chloride ion penetrability into concrete in concrete structures in a severe saline environment. *JSCE Journal of Materials, Concrete Structures and Pavements, 61*, 251–264 (in Japanese).

Cornell, C. A. (1968). Engineering seismic risk analysis. *Bulletin of the Seismological Society of America, 58*, 1583–1606.

Dousti, A., Moradian, M., Taheri, S. R., Rashetnia, R., & Shekarchi, M. (2013). Corrosion assessment of RC deck in a jetty structure damaged by chloride attack. *Journal of Performance of Constructed Facilities, 27*, 519–528.

Ellingwood, B. R. (2005). Risk-informed condition assessment of civil infrastructure: State of practice and research issues. *Structure and Infrastructure Engineering, 1*, 7–18.

Enright, M. P., & Frangopol, D. M. (1999). Condition prediction of deteriorating concrete bridges using Bayesian updating. *Journal of Structural Engineering, 125*, 1118–1125.

Estes, A. C., & Frangopol, D. M. (2003). updating bridge reliability based on bridge management systems visual inspection results. *Journal of Bridge Engineering, 8*, 374–382.

Estes, A. C., Frangopol, D. M., & Foltz, S. D. (2004). Updating reliability of steel miter gates on locks and dams using visual inspection results. *Engineering Structures, 26*, 319–333.

Frangopol, D. M. (2011). Life-cycle performance, management, and optimization of structural safety under uncertainty: Accomplishments and challenges. *Structure and Infrastructure Engineering, 7*, 389–413.

Frangopol, D. M., & Liu, M. (2007). Maintenance and management of civil infrastructure based on condition, safety, optimisation, and life-cycle cost. *Structure and Infrastructure Engineering, 3*, 29–41.

Frangopol, D. M., & Kong, J. S. (2001). Expected maintenance cost of deteriorating civil infrastructures. In D. M. Frangopol & H. Furuta (Eds.), *Life-Cycle Cost Analysis and Design of Civil Infrastructure Systems* (pp. 22–47). Reston, VA: ASCE.

Frangopol, D. M., Saydam, D., & Kim, S. (2012). Maintenance, management, life-cycle design and performance of structures and infrastructures: A brief review. *Structure and Infrastructure Engineering, 8*(1), 1–25.

Frangopol, D. M., Strauss, A., & Kim, S. (2008a). Use of monitoring extreme data for the performance prediction of structures: General approach. *Engineering Structures, 30*, 3644–3653.

Frangopol, D. M., Strauss, A., & Kim, S. (2008b). Bridge reliability assessment based on monitoring. *Journal of Bridge Engineering, 13*, 258–270.

Frangopol, D. M., & Tsompanakis, Y. (Eds.). (2014). *Maintenance and Safety of Aging Infrastructure,* Structures & Infrastructures Book Series, Vol. 10. London: CRC Press / Balkema - Taylor & Francis Group, 746 p.

Japan Society of Civil Engineers (JSCE). (2001). *Standard specifications for concrete structures.* Tokyo: Maruzen.

Kato, E., Iwanami, M., Yokota, H., & Yamaji, T. (2002). Development of life-cycle management system for open type wharf. *Annual Report by Port and Airport Research Institute, 48*, 3–36 (in Japanese).

Komure, K., Hamada, H., Yokota, H., & Yamaji, T. (2002). Development of a model on deterioration progress for RC deck of open type wharf. *Annual Report by Port and Airport Research Institute, 41*, 3–38 (in Japanese).

Li, C. Q. (2003). Life-cycle modeling of corrosion-affected concrete structures: Propagation. *Journal of Structural Engineering, 129*, 753–761.

Li, C. Q. (2004). Reliability based service life prediction of corrosion affected concrete structures. *ASCE Journal of Structural Engineering, 130*, 1570–1577.

Madanat, S. (1993). Optimal infrastructure management decisions under uncertainty. *Transportation Research Part C: Emerging Technologies, 1*, 77–88.

Moradi-Marani, F., Shekarchi, M., Dousti, A., & Mobasher, B. (2010). Investigation of corrosion damage and repair system in a concrete jetty structure. *Journal of Performance of Constructed Facilities, 24*, 294–301.

Mori, Y., & Ellingwood, B. R. (1993). Reliability-based service-life assessment of aging concrete structures. *Journal of Structural Engineering, 119*, 1600–1621.

Okasha, N. M., & Frangopol, D. M. (2009). Lifetime-oriented multi-objective optimization of structural maintenance, considering system reliability, redundancy, and life-cycle cost using GA. *Structural Safety, 31*, 460–474.

Okasha, N. M., Frangopol, D. M., & Orcesi, A. D. (2012). Automated finite element updating using strain data for the life-time reliability assessment of bridges. *Reliability Engineering & System Safety, 99*, 139–150.

Onoufriou, T., & Frangopol, D. M. (2002). Reliability-based inspection optimization of complex structures: A brief retrospective. *Computers & Structures, 80*, 1133–1144.

Saydam, D., Frangopol, D. M., & Dong, Y. (2013). Assessment of risk using bridge element condition ratings. *Journal of Infrastructure Systems, 19*, 252–265.

Taniguchi, O., Tamura, T., Sano, K., & Hamada, H. (2004). Condition state estimate of deteriorated RC superstructure of wharves using inspection results. *Proceedings of Japan Concrete Institute, 26*, 2049–2054 (in Japanese).

Teigen, J. G., Frangopol, D. M., Sture, S., & Felippa, C. A. (1991a). Probabilistic FEM for nonlinear concrete structures. I: Theory. *Journal of Structural Engineering, 117*, 2674–2689.

Teigen, J. G., Frangopol, D. M., Sture, S., & Felippa, C. A. (1991b). Probabilistic FEM for nonlinear concrete structures. II: Applications. *Journal of Structural Engineering, 117*, 2690–2707.

Val, D. V., & Stewart, M. G. (2003). Life-cycle cost analysis of reinforced concrete structures in marine environments. *Structural Safety, 25*, 343–362.

Yoshida, I. (2009). Data assimilation and reliability estimation of existing structure. *COMPDYN 2009 international conference*. Greece: Rhodes.

Mesoscopic simulation of steel rebar corrosion process in concrete and its damage to concrete cover

Airong Chen, Zichao Pan and Rujin Ma

ABSTRACT

In atmospheric or chlorine environment, the rebar corrosion process can be initiated by concrete carbonation or chloride penetration, respectively. Along with the rebar corrosion, tensile stress will occur in concrete around rebars, and cracks will be formed when the stress reaches the tensile strength of concrete. This paper mainly studies the uniform and pitting rebar corrosion processes by lattice model at meso-scale. Firstly, the models of uniform and pitting rebar corrosions are developed. Secondly, an integrated procedure based on the lattice model is proposed to identify the crack pattern in the concrete. Published experimental results are adopted to validate the proposed models and method. Thirdly, some typical cases such as the corrosion process of a single rebar and multi-rebars, rebar corrosion process in the corner and pitting corrosion processes are simulated. Finally, the damage of the rebar corrosion to the concrete cover is studied. The results show that the rebar corrosion in the corner and the pitting corrosion can result in a severer damage to the concrete cover compared with the ordinary uniform rebar corrosion. Furthermore, the corrosion of multi-rebar can result in a larger scale of concrete cover spalling due to the firstly formed transverse crack.

Introduction

The rebar corrosion is one of the major problems which most reinforced concrete (RC) structures and infrastructures may encounter during the service life. As shown in Figure 1, the rebar corrosion process can be divided into three stages. In the first stage, the rebars are protected by the passivation layer around the rebars. However, when the chloride content near the rebars reaches a threshold value, the passivation layer will start to dissolve, which can result in the initiation of the rebar corrosion. In the second stage, the tensile stress can be found in the concrete due to the continuous formation of the corrosion products whose molar volumes are large than that of the iron. When the tensile stress reaches the tensile strength of the concrete, cracks will be formed. In the third stage, the cracks around the rebars will further propagate to form a connected crack pattern. The concrete cover can be damaged due to the cracks and will spall from the entire concrete component (girder, pier, etc.) when the propagation of the cracks reaches a critical extent. Some key time points in these stages, e.g. corrosion initiation, crack formation, cover spalling, are important aspects in the life-cycle design of RC structures.

Numerous efforts have been made to study the rebar corrosion process under different conditions by experiment (Caré & Raharinaivo, 2007; Elsener, 2005; Gadve, Mukherjee, & Malhotra, 2009; Garcés, Andrade, Saez, & Alonso, 2005; Lambert, Page, & Vassie, 1991; Manera, Vennesland, & Bertolini, 2008; Oh & Jang, 2005; Page, Lambert, & Vassie, 1991; Poupard, Aït-Mokhtar, & Dumargue, 2004; Pour-Ghaz, Isgor, & Ghods, 2009; Poursaee & Hansson, 2009; Yu, Shi, Hartt, & Lu, 2010). These experimental researches revealed the mechanism of the rebar corrosion and its consequences on the safety of RC structures either from a macroscopic or microscopic perspective. Corresponding numerical researches were also conducted (Arora et al., 1997; Balafas & Burgoyne, 2010; Bhargava, Ghosh, Mori, & Ramanujam, 2005; Chen & Mahadevan, 2008; Guzmán, Gálvez, & Sancho, 2012; Liu & Weyers, 1998a, 1998b; Meakin, Jøssang, & Feder, 1993; Molina, Alonso, & Andrade, 1993; Shafei, Alipour, & Shinozuka, 2012). These numerical researches can be divided into two categories: (1) analytical models based on the thick-walled cylinder theory; (2) numerical models based on the finite element method and other similar methods. Analytical models are easier to use, but can only calculate the corrosion process of a single rebar. Compared with the analytical approach, numerical models are more flexible and can deal with different cases, e.g. multi-rebar corrosion, corrosion of the rebar in the corner of the RC structures. Most of the above-mentioned numerical researches were conducted at macro-scale, where concrete is considered as a homogeneous material. As a result, the crack pattern obtained in this approach is usually too regular and far from the real situation. The reason for this is that the macroscopic method cannot consider the heterogeneous nature of the concrete which should be modelled as a composite material which consists of aggregates,

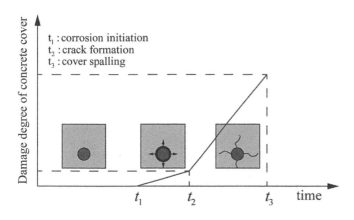

Figure 1. Different stages of rebar corrosion during service life of RC structures.

hardened cement paste and interfacial transition zone (ITZ). The nonuniform distribution of aggregates in the hardened cement paste will absolutely result in the heterogeneity of concrete and an irregular crack pattern as well.

This paper mainly performs a numerical simulation study on the rebar corrosion in the concrete from a mesoscopic perspective. The whole structure of this paper is organised as follows. First of all, the numerical models of the uniform and pitting rebar corrosion processes are developed. Secondly, an integrated procedure based on the lattice model is proposed to identify the crack pattern in the concrete. The published experimental research by Andrade, Alonso, and Molina (1993) is adopted to validate the models and analysis procedure proposed in this paper. Thirdly, some typical examples such as the corrosion process of a single rebar, interaction between corrosion processes of multi-rebars and rebar corrosion process in the corner are simulated. Finally, the fracture process of the concrete cover due to the rebar corrosion is briefly studied. The research in this paper can provide a useful tool to evaluate the deterioration process of RC structures after the initiation of the rebar corrosion and estimate the remaining service life.

Models and methods

Meso-scale model of concrete

At meso-scale, the concrete is usually considered as a composite material which consists of the aggregates, hardened cement paste and ITZ. The hardened cement paste is a homogeneous matrix and the aggregates are inclusions which are randomly placed in the matrix. Thus, the major task to generate the meso-scale model of the concrete is how to properly model the aggregates.

A proper modelling of the aggregate should consider both of the accuracy and the efficiency of the simulation. Although some advanced models of the aggregate were already developed (Garboczi, 2002; Häfner, Eckardt, Luther, & Könke, 2006), a huge amount of computation work may be needed to use these models. Thus, the application of these models in the mesoscopic numerical analysis is strongly restricted by the ability of the computers. This explains well why even in some recently published papers (Ruan & Pan, 2012; Šavija, Pacheco, & Schlangen, 2013), the circle was still used to model the shape of the aggregate because of its simplicity and high efficiency. However, some researches

indicated that the aspect ratio of the aggregate could affect the chloride diffusivity of the concrete (Pan, Ruan, & Chen, 2014), while the circular aggregate cannot reflect this difference. Thus, the modelling of aggregates by circle may be oversimplified in some situations. With this consideration, the shape of the aggregate in this paper is simplified as a random polygon, which is especially suitable to simulate the crushed stones.

The detailed method to use the polygon to generate the meso-scale model of the concrete can be found in (Pan, Chen, & Ruan, 2015; Pan, Ruan, & Chen, 2015). Thus, only a brief introduction is included here. The general idea of this method is to firstly generate a reference circle based on the given aggregate gradation. After that, the reference circle is modified by cutting, stretching and scaling to form the required polygon. The aspect ratio is adopted as the major parameter controlling the shape of the polygon. When the shapes of all polygons are formed, the algorithm called 'taken-and-place' method is used to pack the aggregates (Wang, Kwan, & Chan, 1999). During this procedure, a separation check is conducted for each aggregate to make sure that no aggregates are overlapped with one another in the hardened cement paste. The rebars are explicitly placed in the meso-scale model to consider the influence of the rebars on the packing of aggregates. Figure 2 shows a typical example of the meso-scale model of the concrete generated by the above method.

Corrosion model

Corrosion products

The rebar corrosion process contains very complicated chemicals. Depending on the content of the water and oxygen, as well as the environmental conditions such as the temperature and relative humidity, different products may co-exist during the rebar corrosion process. Common corrosion products and their relative volume ratios compared with the iron are listed as follows (Jaffer & Hansson, 2009): (1) $FeO = 1.7$; (2) $Fe_3O_4 = 2.0$; (3) $Fe_2O_3 = 2.1$; (4) $Fe(OH)_2 = 3.6$; (5) $Fe(OH)_3 = 4.0$; (6) $Fe(OH)_3 3H_2O = 6.2$. Currently, it is still very hard to predict the proportion of each product during the rebar corrosion process. But in a condition with a sufficient supply of the oxygen, $Fe(OH)_2$ and $Fe(OH)_3$ will be the two major products. However, $Fe(OH)_2$ is an extremely unstable product, and can be easily transformed into $Fe(OH)_3$. Therefore, it is assumed in this paper that the corrosion product only consists of $Fe(OH)_3$.

Corrosion rate

Based on the Faraday's law, the mass loss of the rebar due to the corrosion can be calculated as:

$$\Delta m_s = \left(\frac{Q}{F}\right)\left(\frac{M_s}{z}\right) \qquad (1)$$

where Δm_s is the mass loss of the rebar; Q is the total electric charge passed through the rebar; F is the Faraday constant taken as 96,485 C/mol; M_s is the molar mass of the iron; z is the valency which represents the electrons transferred per ion. With the assumption that the corrosion product only consists of $Fe(OH)_3$, z should be 3.0.

In Equation (1), the total electric charge Q can be calculated as follows:

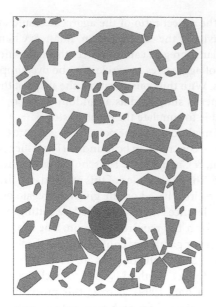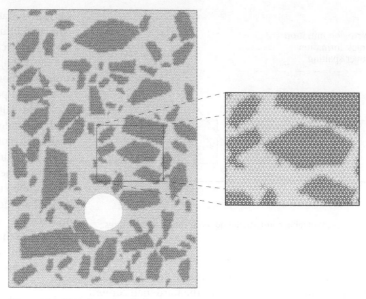

Meso-scale model of concrete   Lattice model

Figure 2. A schematic of discretising mesoscopic model by beam element.

Table 1. Material properties of major corrosion products.

Corrosion products	Fe	FeO	Fe_3O_4	Fe_2O_3	$Fe(OH)_2$	$Fe(OH)_3$	$Fe(OH)_3\ 3H_2O$
Volume ratio	1.0	1.7	2.0	2.1	3.6	4.0	6.2
Molar mass (g/mol)	56	72	232	160	90	107	161
Density (10^3 kg/m³)	7.85	5.93	16.24	10.67	3.50	3.75	3.64

$$Q = i_{corr} \cdot \pi \cdot D_s(t) \cdot L \cdot \Delta t \qquad (2)$$

where i_{corr} is the corrosion current density; $D_s(t)$ is the diameter of the rebar; L is the length of the rebar. Since only a two-dimensional analysis is conducted in this paper, a unit length of the rebar is adopted, i.e. $L = 1.0$ m. During the corrosion process, the diameter of the rebar will diminish due to the mass loss. Thus, D_s is a time-dependent variable and can be calculated by:

$$D_s(t + \Delta t) = \sqrt{D_s^2(t) - \frac{4 \cdot \Delta V_s(t)}{\pi \cdot L}} \qquad (3)$$

where ΔV_s is the diminished volume of the rebar.

Rust expansion

In a uniform corrosion process which is schematically shown in Figure 3, the diminished volume of the rebar should be:

$$\Delta V_s = \frac{\Delta m_s}{\rho_s} \qquad (4)$$

where ρ_s is the density of the iron; Δm_s is determined by Equation (1). Accordingly, the increased volume of the corrosion product can be calculated as:

$$\Delta V_r = \Delta V_s \cdot \frac{\rho_s}{\rho_r} \cdot \frac{M_r}{M_s} = \Delta V_s \cdot \frac{r_\rho}{r_M} \qquad (5)$$

where r_ρ and r_M are the density and molar mass ratios of the iron to the corrosion product. In this paper, it is assumed that

the corrosion product only consists of $Fe(OH)_3$. So $r_\rho = 2.093$ and $r_M = .5234$ (these values are calculated based on the data in Table 1).

Based on Equations (4) and (5), the diameter of the corroded rebar including the corrosion product can be calculated by:

$$D_r(t) = \sqrt{D_0^2 + \frac{4 \cdot \int_0^t (\Delta V_r - \Delta V_s)dt}{\pi \cdot L}} \qquad (6)$$

where D_0 is the initial diameter of the rebar. Thus, the thickness of the rust layer can be calculated as:

$$t_r(t) = \frac{D_r(t) - D_s(t)}{2} \qquad (7)$$

where $D_s(t)$ can be determined by Equation (3).

It should be noted that D_r in Equation (6) does not consider the deformations of the corrosion product itself due to the restriction of the surrounding hardened cement paste against the expansion. This simplification actually assumes that the corrosion product has a much larger Young's modulus than the hardened cement paste. In the early stage of the rebar corrosion, the corrosion product will occupy the porous zone around the rebar. Thus, a free expansion of corrosion product exists during the corrosion process (Michel, Pease, Geiker, Stang, & Olesen, 2011; Wong, Zhao, Karimi, Buenfeld, & Jin, 2010). To consider this, $t_r(t)$ in Equation (7) can be modified as follows:

$$t_r^*(t) = t_r(t) - t_{buff} \qquad (8)$$

where t_{buff} is the thickness of the porous zone, which serves as a buffer for the rust expansion. The value of t_{buff} observed in the

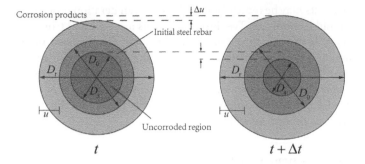

Figure 3. A schematic of rust expansion in uniform rebar corrosion.

experiments shows a large deviation, and was reported in the range of 10~100 μm based on the microstructure of the concrete (Liu & Weyers, 1998b).

Compared with the uniform corrosion, the pitting corrosion is more common in reality, especially when the corrosion is induced by the chloride penetration process, during which the chloride ions can accumulate in a small area on the surface of the rebar. To simulate the pitting corrosion process, the assumed types of the corrosion distribution in Jang and Oh (2010) are adopted, as shown in Figure 4. Based on this assumption, the thickness of the rust layer in a pitting corrosion can be determined by the following procedure:

(1) Calculate ΔV_r and ΔV_s based on the Faraday's law. In a pitting corrosion, only a part of the surface of the rebar can have chemical reactions.
(2) For an equivalent uniform corrosion with the same ΔV_r and ΔV_s determined in the first step, calculate the thickness of the rust layer by Equation (7).
(3) The thickness of the rust layer in a pitting corrosion can be calculated as:

$$t_{r,p}(t,\gamma) = \alpha \cdot t_r(t) \cdot \left(1.0 - \frac{\gamma}{\gamma_m}\right) \tag{9}$$

where α is defined as the ratio of the depth of pitting corrosion to that of uniform corrosion; t_r is the thickness of the rust layer in the equivalent uniform corrosion. The meanings of γ and γ_m are shown in Figure 5. For the three types of corrosion distribution in Figure 4, the values of the parameters α and γ_m should be $\alpha = 2.0$, 4.0, 8.0 and $\gamma_m = \pi$, $\pi/2$, $\pi/4$, respectively.

Mechanical model

To simulate the fracture process of the concrete due to the rebar corrosion, the lattice type models are used in this paper. These models were initially proposed to study the fracture mechanisms in heterogeneous materials (Moukarzel & Herrmann, 1992). So they can also be used to simulate the concrete fracture process. In these models, the material is spatially discretised into a series of small truss or beam elements which can transfer the forces, as shown in Figure 2. Each individual element exhibits a linear elastic behaviour. The fracture process can be simulated by performing a linear elastic analysis and removing the failed elements from the model. This analysis is repeated several times and the failed elements are removed continuously. Thus, the crack pattern in the concrete can be finally obtained. Since the heterogeneous property of the concrete is explicitly considered in the lattice model, the crack pattern will be more realistic than that obtained in a homogeneous approach. It should be also mentioned that although all the elements in the lattice models are assumed to have a linear elastic property, the overall concrete model can still exhibit a nonlinear behaviour and the ductile response can be also achieved (Schlangen, 1993). More details about the lattice model and its application in the simulation of the concrete fracture process can be found in some publications (Schlangen, 1993; Schlangen & Garboczi, 1997).

Since the element in the lattice model only exhibits a linear elastic behaviour, the failure criteria for each element can be defined as:

$$\sigma_t > f_t \tag{10}$$

where σ_t and f_t are the tensile stress and strength of the element, respectively. σ_t in the above equation can be calculated as:

$$\sigma_t = \alpha_N \frac{N}{A} + \alpha_M \frac{\max(M_i, M_j)}{W} \tag{11}$$

where N is the axial force; A is the area of the cross section of the element; M_i and M_j are bending moments at the node i and j, respectively; W is the section modulus; α_N and α_M are two parameters representing the contributions of the axial force and bending moment on the tensile stress of the element. A detailed discussion on these two parameters can be found elsewhere (Schlangen, 1993). Based on these researches, the values of $\alpha_N = 1.0$ and $\alpha_M = 0.05$ are commonly adopted, and also be used in this paper.

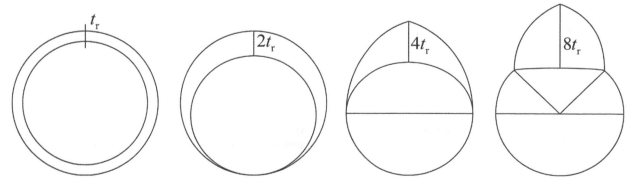

Figure 4. Different types of corrosion distribution in pitting corrosion (Jang & Oh, 2010).

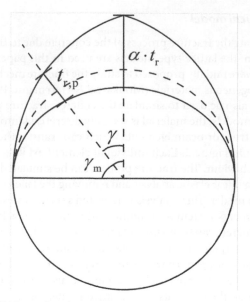

Figure 5. Thickness of rust layer in pitting corrosion.

Based on the above discussion, an integrated algorithm to simulate the processes of the rebar corrosion and concrete fracture by the lattice model is proposed and shown in Figure 6. Compared with some current lattice models (Qian, 2012; Schlangen & Garboczi, 1997; Schlangen & Van Mier, 1992), the most important improvement of the algorithm in this paper is to implement an incremental analysis in the lattice model. Firstly, the radial displacement is exerted on the concrete surrounding the rebar step by step, based on the incremental time step Δt. Secondly, in each time step, all the elements which satisfy the failure criteria in Equation (10) will be removed from the model. Due to the removal of failed elements, the tensile stresses of the remaining

elements may be redistributed. Therefore, an iteration procedure should be adopted. Two kinds of results are expected from the analysis in each time step. If no elements are removed in an iteration step, the whole analysis procedure advances to the next time step. If more and more elements failed and the maximum displacement of the model reaches a permitted value, the entire model should be considered as completely failed, and the whole analysis procedure will stop.

Finite element model

Finite element mesh and heterogeneity of concrete

When the meso-scale model of the concrete is generated, the rebars are subtracted from the model. After that, the remaining part of the model is projected to a background mesh which consists of regular triangles. Each side of the triangle in the mesh represents one beam element. To implement the heterogeneity of concrete, the type and the material property (Young's modulus, tensile strength, etc.) of each beam element are determined according to its position in the meso-scale model, as schematically shown in Figure 2. When the entire beam element lies in the hardened cement paste or the same aggregate, the mechanical property of the hardened cement paste or aggregate is attributed to this element, respectively. If the beam element crosses the edge of an aggregate, it is an interface element. The mechanical property of such element is determined as:

$$E_{eq} = \frac{l}{\frac{l_a}{E_a} + \frac{l_c}{E_c} + \frac{l_i}{E_i}} \qquad (12)$$

$$f_t^{eq} = \min\{f_t^a, f_t^c, f_t^i\} \qquad (13)$$

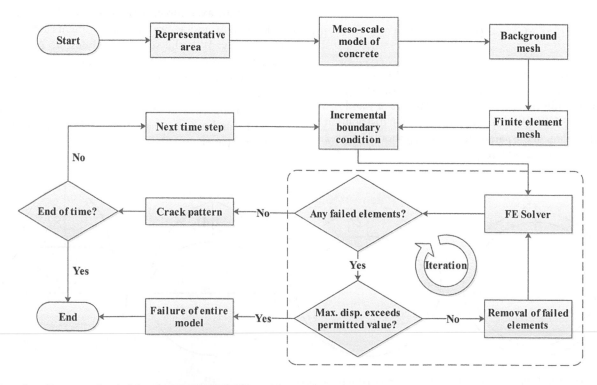

Figure 6. Procedure of incremental analysis for rebar corrosion by lattice model.

Figure 7. Equivalent Young's modulus of interface element in lattice model.

Table 2. Mechanical properties of hardened cement paste, aggregate and ITZ used in lattice model.

Type of constituent	Young's modulus (GPa)	Tensile strength (MPa)
Aggregate	70	8
Hardened cement paste (or mortar)	25	4
ITZ	15	2.5

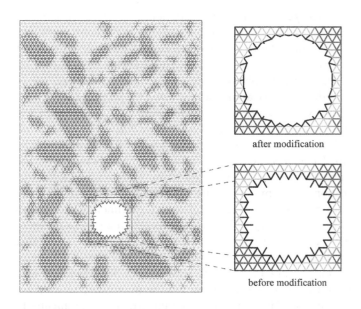

after modification

before modification

Figure 8. Modification of mesh near surface of rebar.

where E_a, E_c and E_i are the elastic moduli of the aggregate, hardened cement paste and ITZ, respectively; f_t^a, f_t^c and f_t^i are the tensile strengths of the aggregate, hardened cement paste and ITZ, respectively; l_a, l_c and l_i are the lengths of the parts of the element inside the aggregate, hardened cement paste and ITZ, respectively (see Figure 7); l is the length of the element; E_{eq} and f_t^{eq} are the Young's modulus and tensile strength of the interface element. To use Equation (12), the thickness of the ITZ should be assumed. According to some published experimental researches (Ollivier, Maso, & Bourdette, 1995; Scrivener & Nemati, 1996), the thickness of the ITZ is usually in the range of $10 \sim 50\,\mu m$. Thus, $l_i = 30\,\mu m$ is adopted for all the interface elements in this paper.

The general idea of Equations (12) and (13) is to homogenise the interface element which consists of different constituents, i.e. aggregate, hardened cement paste, ITZ, into an equivalent element. The Young's modulus of the equivalent element is obtained based on the assumption that the elements before and after the homogenisation should have the same axial strain under the same axial external force. The tensile strength of the equivalent element is determined by the weakest constituent, and thus should be the minimum among the tensile strengths of the aggregate, hardened cement paste and ITZ.

The mechanical properties of the aggregate, hardened cement paste and ITZ used in this paper are listed in Table 2. It can be seen that the ITZ has the smallest tensile strength. According to Equation (13), the interface element will have the same tensile strength as the ITZ. As a result, the crack can be expected to initiate and propagate in these elements. In other words, the crack pattern is strongly dependent on the distribution of the aggregates in the hardened cement paste, since the interface elements only exist at the edges of the aggregates.

During the mesh procedure, there may be some beam elements across the edge of the rebars. These elements need to be modified according to the following procedure:

(1) Find the interface elements across the edge of the rebar.
(2) Find the nodes which belong to the interface elements found in Step (1) and are also inside the rebar.
(3) Change the positions of the nodes found in Step (2) to let these nodes lie on the edge of the rebar with an equal interval spacing between the adjacent nodes.

A typical example by following the above procedure is shown in Figure 8.

To ensure that the lattice model can realistically reflect Poisson's ratio of the concrete, the length of the beam element and the height of the cross section should satisfy the following equation (Schlangen & Garboczi, 1997):

$$v = \frac{1 - (\frac{h}{l})^2}{3 + (\frac{h}{l})^2} \tag{14}$$

where v is Poisson's ratio of the concrete, l is the length of the beam element, h is the height of the cross section of the beam element. In the simulation, l is firstly determined based on the minimum size of the aggregates. Subsequently, h can be calculated according to Equation (14).

Displacement boundary conditions

To express the effect of the rust expansion on the concrete in the finite element analysis, two types of boundary conditions can be used, i.e. displacement and force caused by the restriction of the concrete on the rust expansion of the rebar. The former one is adopted in this paper, as it is more convenient to calculate the displacement based on the models of the rust expansion, e.g. Equations (7) and (9).

In a uniform corrosion process, the radial displacement boundary condition used in an incremental analysis after the consideration of t_{buff} can be expressed as:

$$\Delta u^*(t + \Delta t) = \begin{cases} 0 & u(t + \Delta t) < t_{buff} \\ u(t + \Delta t) - u(t) & u(t + \Delta t) \geq t_{buff} \end{cases} \tag{15}$$

where $u(t)$ is calculated by:

$$u(t) = t_r - \frac{\Delta D_s}{2} = t_r - \frac{D_0 - D_s(t)}{2} \tag{16}$$

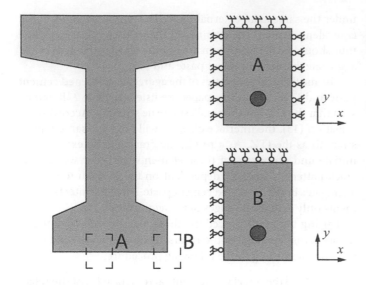

Figure 9. Displacement boundary conditions on surfaces of meso-scale model for rebar corrosion in normal region and corner.

According to this boundary condition, the parameter t_{buff} can delay the initiations of the stress and cracks.

In a pitting corrosion process, the situation is much more complicated. $t_{r,p}(t, \gamma)$ in Equation (9) only gives the thickness of the rust layer. According to Equation (16), the decrease in the rebar diameter in a pitting corrosion ($\Delta D_{s,p}$) should be also obtained to calculate the displacement boundary condition. Similar to $t_{r,p}(t, \gamma)$, it is also assumed that $\Delta D_{s,p}$ follows a linear decrease with the angle γ in Figure 5. Thus, the displacement boundary condition in a pitting corrosion process can be expressed as:

$$
\begin{aligned}
u_p(t, \gamma) &= t_{r,p} - \frac{\Delta D_{s,p}(t)}{2} \\
&= \left[t_r - \frac{\Delta D_s(t)}{2} \right] \cdot \alpha \cdot f(\gamma) = u(t) \cdot \alpha \cdot f(\gamma)
\end{aligned}
\tag{17}
$$

where $f(\gamma) = 1 - \gamma/\gamma_m$; $u(t)$ is the displacement in an equivalent uniform corrosion with the same ΔV_s and ΔV_r in the pitting corrosion. When $u_p(t, \gamma)$ is calculated, Equation (15) can be also used to determine the displacement boundary condition in a pitting corrosion process.

Apart from the displacement boundary conditions around the rebar, some of the displacements on the surface of the meso-scale model should be also fixed to avoid the drift of the entire model. To decrease the amount of the computational work, the model used in a mesoscopic analysis is often a small region on the cross section of the full-scale component, as shown in Figure 9. For the rebar corrosion in region A, the deformation of the model will be restricted by the adjacent regions. Thus, both of the displacements in x and y directions are fixed on the two sides of the model. For the rebar corrosion in region B which is in the corner of the cross section, only the displacements in x and y directions on one side are fixed.

Results and discussion

Experimental validation

To validate the mesoscopic method proposed in this paper, the experimental data in Andrade et al. (1993) were adopted. In their research, the artificial accelerated rebar corrosion processes in three different concrete specimens were studied. As shown in Figure 10, the first specimen was to study the rebar corrosion in the corner, while the other two specimens were to evaluate the influence of the thickness of the concrete cover on the rebar corrosion. The size of the specimens used in the experiment was 15 cm × 15 cm × 38 cm, while the diameter of the embedded rebar was 16 mm. To conduct a 2D simulation, the cross section of the specimen (15 cm × 15 cm) is adopted. The electrical current to induce the rebar corrosion was applied to the specimen artificially. To decrease the time period of the experiment, the current density was set as $100\,\mu A/cm^2$, which could efficiently accelerate the development of the crack on the surface of the specimen only in several days. More detailed information on the experiment can be found in the reference (Andrade et al., 1993).

Based on the above description of the experiment, corresponding numerical simulations using the proposed incremental analysis method by the lattice model were conducted. Since the specimens were just put on the table during the experiment, only the displacement in y direction on the surface of the model which contacts with the table is fixed, as shown in Figure 10. The mesh size, i.e. the length of the beam element used in the simulation, is .5 mm. The results, compared with the experimental ones, are shown in Figure 11. For each specimen, several values of t_{buff} in Equation (8) were tried to obtain a good agreement between the experimental and simulation results. It is found that the best values of t_{buff} are different for these three specimens, i.e. for Specimen I, II and III, respectively. This deviation may be attributed to the differences in the microstructure of these specimens. This result shows the uncertainty and difficulty to determine the proper value of t_{buff}. With this consideration, t_{buff} can be neglected in the practical engineering as a safer approach.

It should also be noticed that the experiment conducted in a laboratory is often accelerated to shorten the period, which is different from the real situation of the rebar corrosion in a natural condition. As a result, the proposed method in this paper needs to be further validated by the in-situ test of the rebar corrosion in real structures, which will be conducted in the future.

Uniform corrosion

After the validation of the proposed numerical method, the uniform corrosion of a single rebar is firstly studied. The model used here is 100 mm × 150 mm, including a rebar with the diameter of 20 mm. Three different thicknesses of the concrete cover, i.e. $d_c = 20, 30, 40$ mm, are adopted. The current density is $1.0\,\mu A/cm^2$ which represents a moderate corrosion rate in the nature (Chen & Mahadevan, 2008). The mesh size is 0.5 mm.

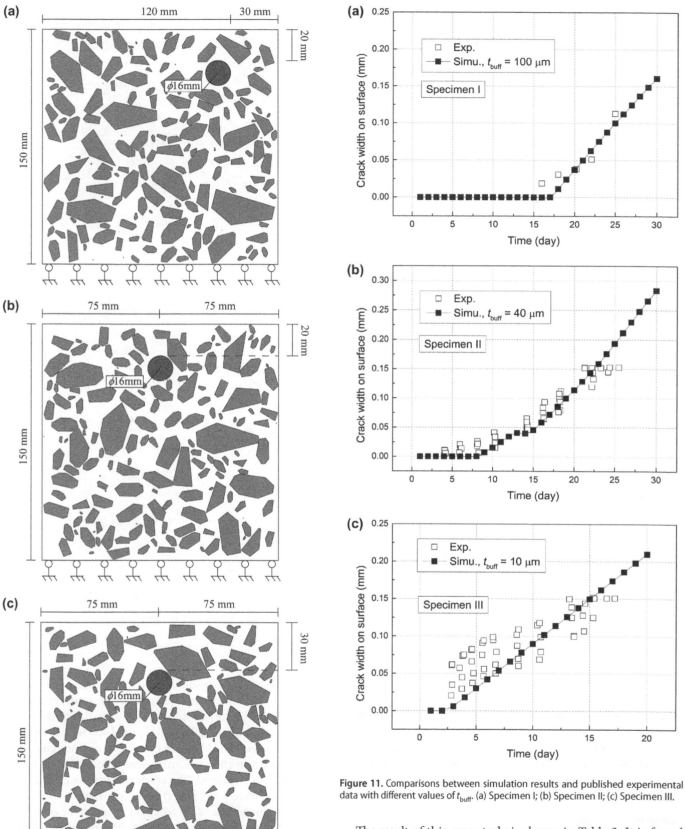

Figure 10. A schematic of meso-scale models and displacement boundary conditions used in experimental validation. (a) Specimen I; (b) Specimen II; (c) Specimen III.

Figure 11. Comparisons between simulation results and published experimental data with different values of t_{buff}. (a) Specimen I; (b) Specimen II; (c) Specimen III.

The result of this case study is shown in Table 3. It is found that when the thickness of the concrete cover is small, a connected major crack can be found from the rebar to the exposed surface of the model. The position and shape of this major crack are dependent on the distribution of the aggregates in the hardened cement paste. When the thickness of the concrete cover increases gradually, an inverted 'V' shaped crack pattern will

Table 3. Crack patterns and deformed shape of model caused by uniform rebar corrosion at the time of 20 years.

d_c	$w_{cr} > 50$ μm	$w_{cr} > 100$ μm	$w_{cr} > 200$ μm	Deformed shape

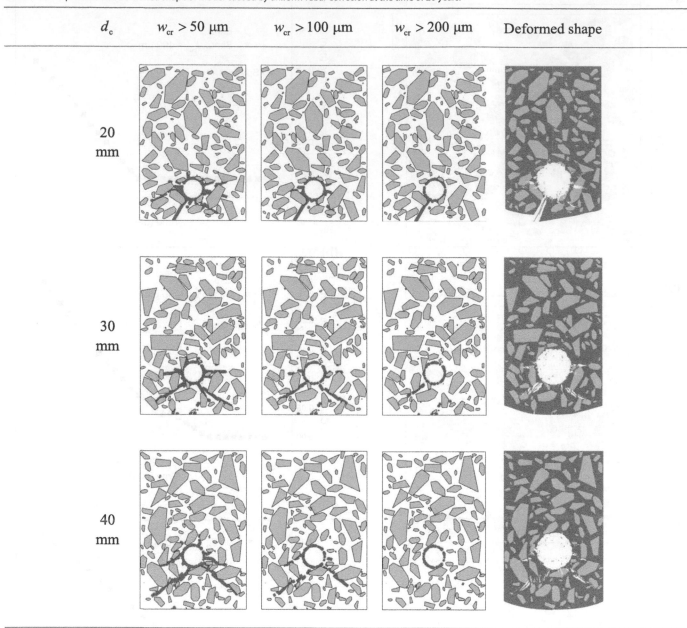

be more likely to develop in the model and the cracks are more difficult to propagate to the exposed surface of the model. Besides the crack pattern, the displacements of the entire model in these three cases are also shown in Table 3. With the purpose of a clear plot, these displacements are amplified by 20 times. These results can provide a direct judgement and understanding when and how the spalling of the concrete cover will happen.

Corrosion in corner

In a chlorine environment, the rebar in the corner of the concrete component will be attacked by the chloride penetration from two directions at the same time. As a result, the initiation of the rebar corrosion will be advanced compared with that of the other rebars. Thus, it is important to investigate the corrosion process of the rebar in the corner.

Table 4 shows a case study on this topic. The model used here is the same as that in the case study of the uniform corrosion. As a rebar in the corner, it has two thicknesses of the concrete cover corresponding to the two exposed surfaces. Both of these thicknesses are set as 30 mm in the simulation. As a comparison, the corrosion in the same meso-scale model but in a normal region is also simulated. The crack patterns and the deformed shapes (amplified by 20 times) in these two cases are compared in Table 4. In the case of the rebar corrosion in the corner, it can be seen that a major crack with the width larger than 200 μm has been already formed and propagated to the left surface of the model at the time of 20 years. Furthermore, a part of the concrete cover already started to spall from the model according to the deformed shape. However, for the corresponding rebar corrosion in the normal region, the concrete fracture process is relatively much slower, and neither of the above results can be found.

Table 4. A comparison of crack patterns and deformed shape of model between uniform rebar corrosion in normal region and corner at the time of 20 years.

	$w_{cr} > 50\ \mu m$	$w_{cr} > 100\ \mu m$	$w_{cr} > 200\ \mu m$	Deformed shape

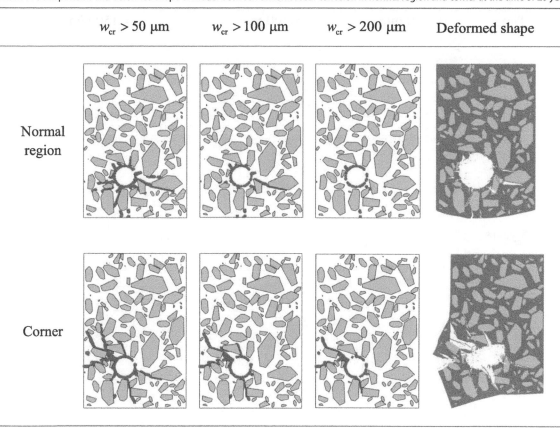

Normal region
Corner

The main reason for the differences between the results of the rebar corrosion in the corner and the normal region is due to the difference in the boundary condition used in the simulation. As shown in Figure 9, the rebar corrosion in the normal region is restricted by the adjacent regions on both sides, while the rebar corrosion in the corner is only hindered by one side. As a result, the crack formed around the rebar in the corner is more likely and easier to propagate to the free surface, i.e. the left side of the model in this case study.

Based on the above discussion, the corner of the concrete component often undertakes a more rapid deterioration process than the normal region. Not only the initiation of the rebar corrosion can be advanced due to the simultaneous chloride penetration from two directions, but also the processes of the crack propagation and the spalling of the concrete cover can be accelerated according to the case study in this section. Therefore, the rebars embedded in the corner of the concrete component should be more carefully and frequently inspected. In some severe environment, some special rebars such as the epoxy rebar and stainless rebar may be required.

Corrosion of multi-rebar

The case study of the uniform corrosion is only applicable to the isolated rebar. If the adjacent rebars corrode at the same time, the corrosion processes of these rebars can have interactions with each other, which will be investigated in this section. The mode used here is $(100 + d)$ mm $\times$ 150 mm, where d is the distance between the centres of the rebars. Two rebars are embedded at (50 mm, 40 mm) and $(50 + d$ mm, 40 mm) in the model. The

other parameters are the same as those in the case study of the uniform corrosion. Three different values of d, i.e. $d = 50$, 100 and 150 mm, are adopted in the simulations. The cracks whose widths are larger than 200 μm, as well as the deformed shape (amplified by 20 times) at the time of 20 years, are shown in Table 5. It is found that along with the development of the rebar corrosion, the cracks between the rebars can be connected with one another. Therefore, before the cracks propagate to the exposed surface of the model and become visible or detectable, a transverse major crack connecting all the rebars which corrode at the same time may already exist inside the concrete. In this situation, a large part of the concrete cover may spall from the model in a sudden at the same time, which is more hazardous to the durability and safety of RC structures, compared with the corrosion of a single rebar.

Pitting corrosion

All the three types of pitting corrosions shown in Figure 4 are investigated. To compare with the uniform corrosion, the same meso-scale model in Table 3 ($d_c = 40$ mm) is adopted. The displacement boundary conditions used in the simulation are determined according to Equation (17). By following the same analysis procedure, the crack patterns and the deformed shape (amplified by 20 times) of the model caused by the pitting corrosion of the rebar are obtained and shown in Table 6. Since only a part of the rebar surface can have chemical reactions, the loss of the cross section of the rebar in a pitting corrosion is smaller than that in a uniform corrosion. However, the results show that the pitting corrosion can result in a much severer damage to the concrete

Table 5. Crack patterns of model caused by corrosion of multi-rebar with different distances between rebars.

Time	$d = 50$ mm	$d = 100$ mm	$d = 150$ mm
10 years	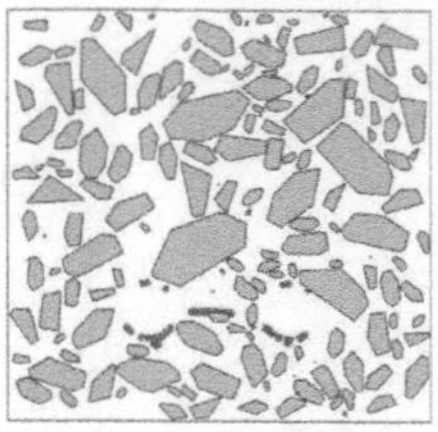	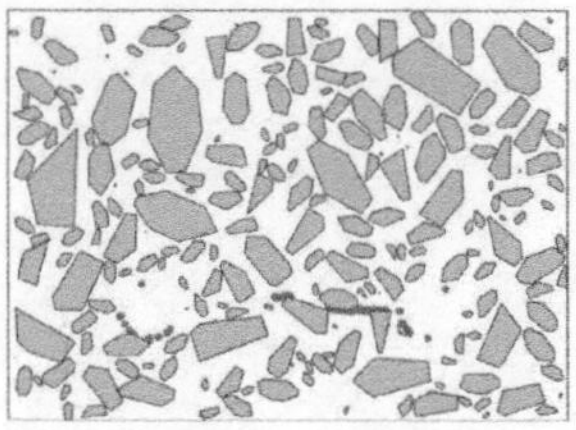	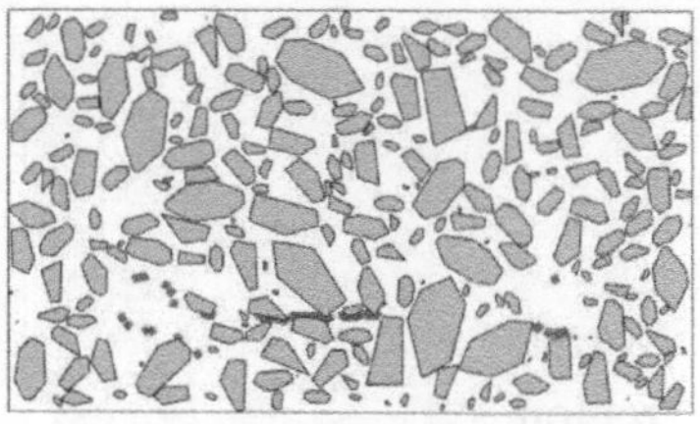
15 years	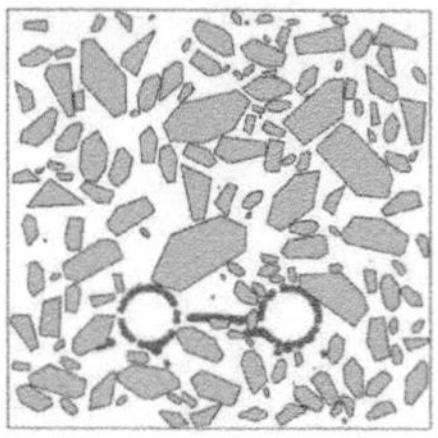	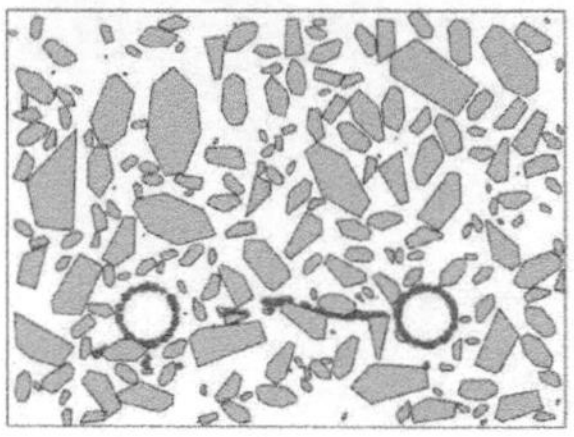	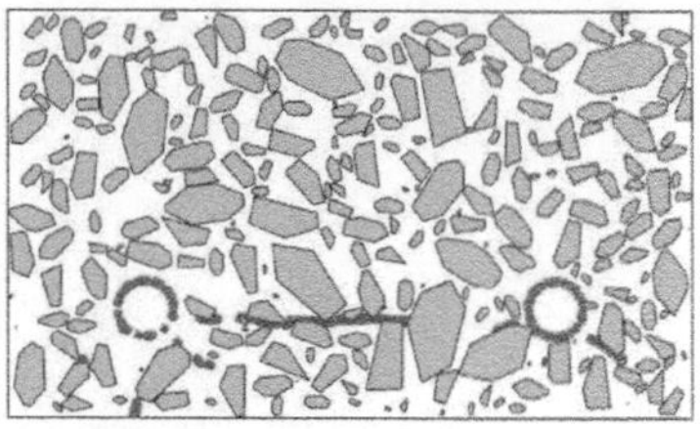
20 years	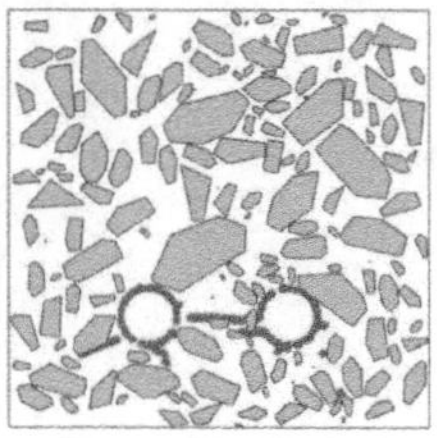	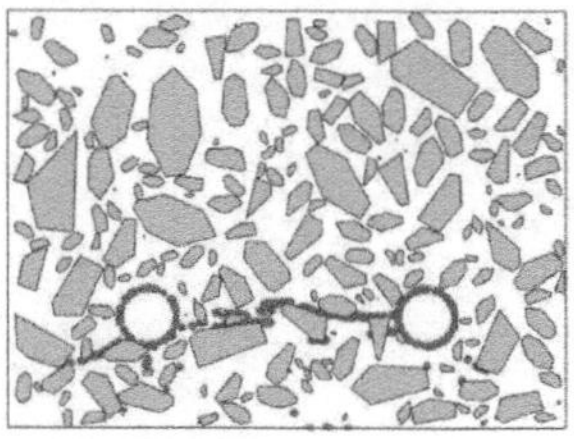	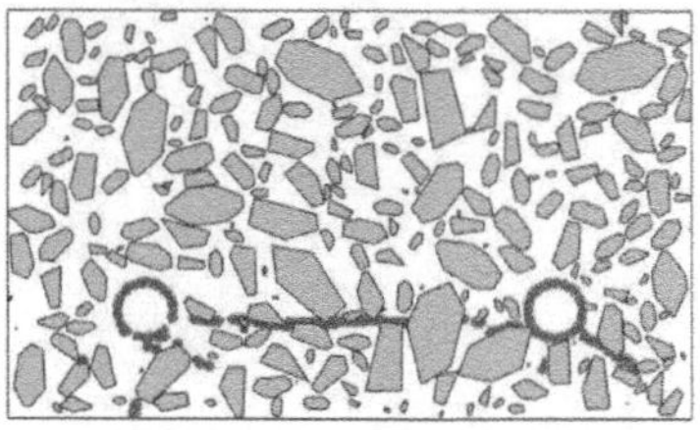
Deformed shape	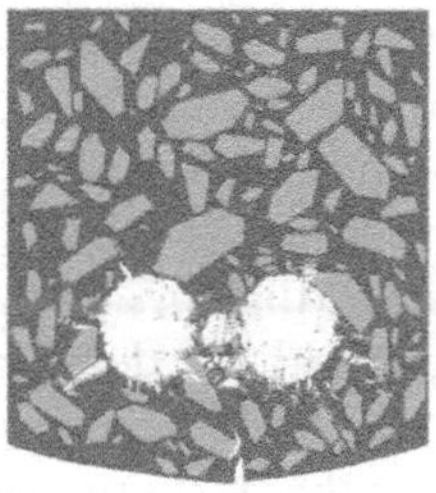	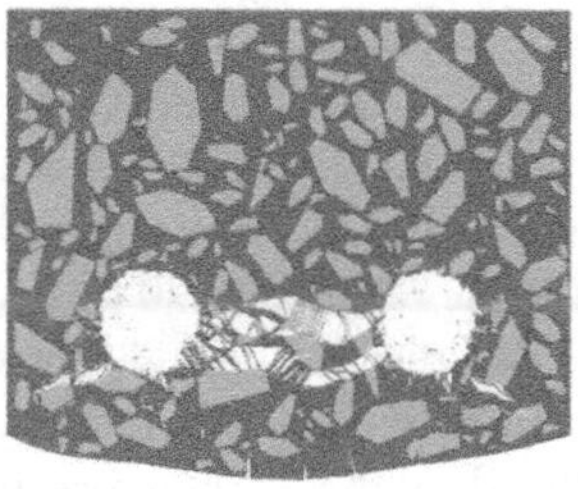	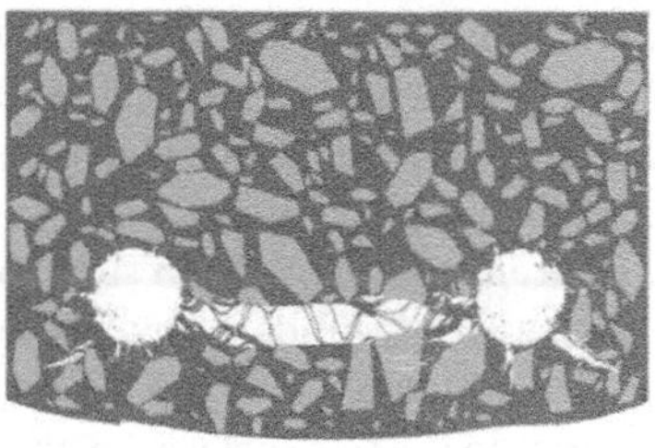

cover than the uniform corrosion due to the differences in the rust distribution.

Currently, the experimental researches on the concrete cracking due to the pitting corrosion are still insufficient. The main reason is that the pitting corrosion process is difficult to simulate in a laboratory due to the long duration of the experiment. Thus, the simulated results of the pitting corrosion cannot be quantitatively validated yet. However, these results are obtained based on the assumed types of the rust distribution shown in Figure 4 which have been already confirmed in the experiment (González, Andrade, Alonso, & Feliu, 1995). In this sense, the case study in this section can still provide a reliable qualitative evaluation on the damage of the pitting corrosion to the concrete cover.

Summary and discussion

Based on the above case studies, some typical kinds of the spalling of the concrete cover induced by different types of the

rebar corrosion are identified and summarised in Figure 12. The results of the crack pattern and deformed shape also provide a perceptual knowledge on the deterioration process of the concrete cover due to the rebar corrosion.

To more accurately discuss the influence of the rebar corrosion on the concrete cover, a proper index should be defined firstly to represent the damage level of the concrete cover. In the current lattice model, all the failed elements are removed from the model step by step. Therefore, a straightforward idea is to use the percentage of the failed elements to define the damage level of the concrete cover, which is shown below:

$$\lambda = \frac{N_{fe}}{N_e} \tag{18}$$

where N_{fe} and N_e are the numbers of the failed elements and all the elements in the concrete cover, respectively. Figure 13 shows the results of λ in the case study of the pitting corrosion in Table 6. As can be seen, only the cases of $\alpha = 4$ and $\alpha = 8$

Table 6. Crack patterns and deformed shape of model caused by pitting corrosion with different types of rust distribution at the time of 20 years.

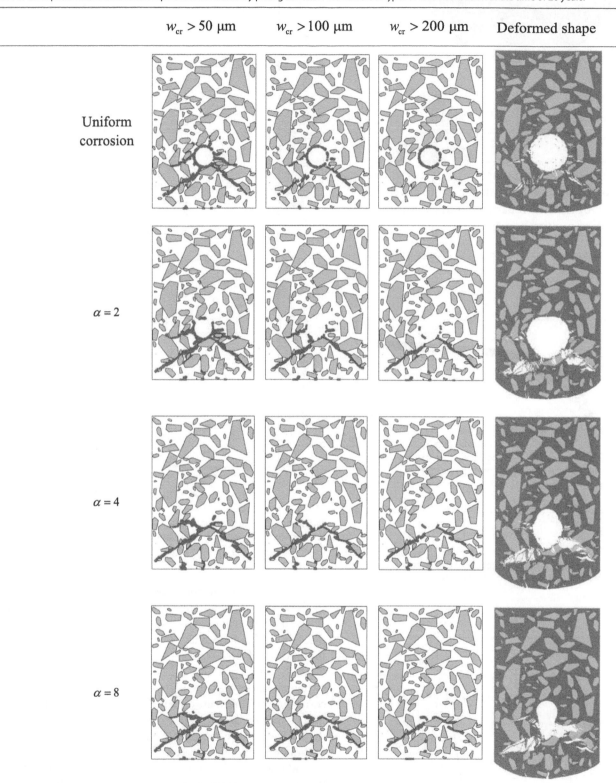

	$w_{cr} > 50\ \mu m$	$w_{cr} > 100\ \mu m$	$w_{cr} > 200\ \mu m$	Deformed shape
Uniform corrosion				
$\alpha = 2$				
$\alpha = 4$				
$\alpha = 8$				

show a slightly larger λ than the other cases. This result indicates that λ cannot be used alone to represent the damage level of the concrete cover.

Another idea is to adopt the probability density function (PDF) of the crack width of the failed elements. This function can provide the information on the percentages of different crack widths of all failed elements. The PDF here is obtained based on the following procedure. Firstly, the base 10 logarithm of the crack width of the ith failed element in the concrete cover is defined as:

$$X_i = \log_{10}(w_{cr}^i) \quad i = 1, 2, 3, \ldots, N_{fe} \tag{19}$$

Secondly, based on the calculated X_i, the function *ksdensity* in MATLAB is used to estimate the PDF.

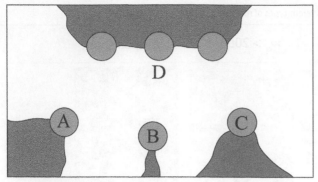

A: rebar corrosion near a corner
B: corrosion of isolated rebar under a thin cover
C: corrosion of isolated rebar uner a thick cover
D: corrosion of multi-rebars

Figure 12. Typical types of spalling of concrete cover caused by different types of rebar corrosion.

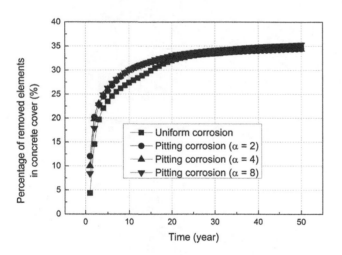

Figure 13. Percentage of failed elements in concrete cover for uniform corrosion and different types of pitting corrosions studied in Table 6.

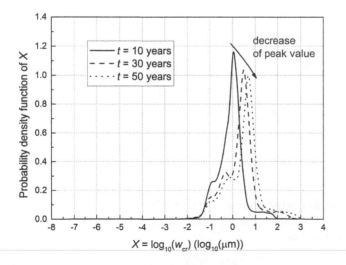

Figure 14. PDF of common logarithm of crack width in concrete cover for uniform corrosion studied in Table 6.

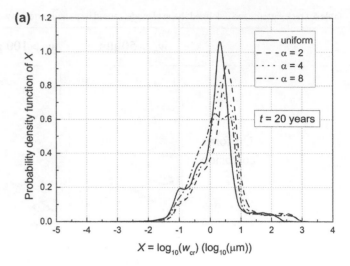

Figure 15. A comparison of PDF of common logarithm of crack width in concrete cover between uniform and pitting corrosions studied in Table 6. (a) $t = 20$ years; (b) $t = 50$ years.

Taking the uniform corrosion in Table 6 as an example, the PDFs of X at different time are plotted in Figure 14. It is clearly shown that the major part of the curve has been shifted towards the right side and the peak value of the function decreases with the time as well. Both of these results indicate that the small cracks inside the model gradually developed into the larger ones. By following the above idea, the PDFs of X in the uniform and pitting corrosions in Table 6 are obtained. Figure 15(a) and (b) plot the PDFs at the time of 20 and 50 years, respectively. As shown, most of the crack widths are between 10^0 and 10^1 μm, no matter whether it is a uniform or pitting corrosion. But differences can be still found around a large crack width ($w_{cr} \geq 10^2$ μm). As indicated, the pitting corrosion can produce more failed elements with a large crack width than the uniform corrosion. This result well explains the differences of the deformed shapes of the model shown in Table 6.

Conclusions

The major purpose of this paper was to advance the understanding of the rebar corrosion process at meso-scale. To implement this purpose, the lattice model was adopted and improved to obtain the realistic crack pattern inside the concrete cover. The related models of the rebar corrosion, rust expansion and mechanical behaviour of the concrete were comprehensively demonstrated. Published experimental results obtained from an accelerated rebar corrosion were introduced to validate the proposed method and models. Several typical case studies on the uniform corrosion of a single rebar, rebar corrosion in the corner, corrosion of multi-rebar and pitting corrosion of a single rebar were conducted as well to illustrate the potential application of the proposed mesoscopic method.

Based on the above achievements, some further researches can be conducted in the future. One of the most important topics is to establish the criterion about the time point of the concrete cover spalling in the framework of the proposed mesoscopic method by the lattice model. In the current research, the deformed shape of the model can already provide a qualitative judgement on the spalling of the concrete cover. But for a more accurate, strict and mathematical definition, deeper researches are absolutely necessary. Based on the further researches and improvements, the proposed mesoscopic method in this paper can provide the researchers and engineers a new insight and tool to estimate the remaining service life of RC structures after the initiation of the rebar corrosion from the perspective of the life-cycle design theory and method.

Nomenclature

Δm_s	mass loss of rebar
Q	total electric charge passed through rebar
F	Faraday constant (96,485 C/mol)
M_s	molar mass of iron
z	valency of corrosion products
i_{corr}	corrosion current density
D_s	diameter of rebar
D_0	initial diameter of rebar before corrosion
D_r	diameter of rebar including rust layer
ΔD_s	diameter change of rebar in a uniform corrosion
$\Delta D_{s,p}$	diameter change of rebar in a pitting corrosion
L	length of rebar
Δt	incremental time step used in analysis
t	time
ΔV_s	volume change of rebar after corrosion
ΔV_r	volume change of corrosion products
ρ_s	density of iron
ρ_r	density of corrosion products
r_ρ	density ratio of iron to corrosion products $(r_\rho = \rho_s/\rho_r)$
r_M	molar mass ratio of iron to corrosion products $(r_M = M_s/M_r)$
u	radial displacement due to uniform corrosion around rebar

u_p	radial displacement due to pitting corrosion around rebar
Δu^*	incremental radial displacement used as boundary condition in analysis after considering t_{buff}
t_r	thickness of rust layer
t_{buff}	thickness of porous layer around rebar
t_r^*	thickness of rust layer after considering t_{buff}
$t_{r,p}$	thickness of rust layer in a pitting corrosion
α	parameter which represents the type of rust distribution in a pitting corrosion
γ	angle from the tip of rust distribution
γ_m	maximum permitted value of γ
E_a	Young's modulus of aggregate
E_c	Young's modulus of hardened cement paste
E_i	Young's modulus of ITZ
f_t^a	tensile strength of aggregate
f_t^c	tensile strength of hardened cement paste
f_t^i	tensile strength of ITZ
l_a	length of part of interface element inside aggregate
l_c	length of part of interface element inside hardened cement paste
l_i	length of part of interface element inside ITZ
E_{eq}	equivalent Young's modulus of interface element
f_t^{eq}	equivalent tensile strength of interface element
σ_t	axial tensile stress of beam element in lattice model
f_t	tensile strength of beam element in lattice model
N	axial force on cross section of beam element in lattice model
M_i, M_j	bending moments at node i and j of beam element in lattice model
A	area of cross section of beam element in lattice model
W	section modulus of beam element in lattice model
α_N, α_M	parameters representing the contributions of axial force and bending moment on tensile stress of beam element in lattice model
v	Poisson's ratio of concrete
h	height of cross section of beam element in lattice model
l	length of beam element in lattice model
d_c	thickness of concrete cover
d	distance between centres of rebar
λ	percentage of failed elements in concrete cover
N_{fe}, N_e	total numbers of the failed elements and all elements in concrete cover
w_{cr}	crack width

Disclosure statement

No potential conflict of interest was reported by the authors.

Funding

This work was supported by the National Natural Science Foundation of China [grant number 50878145], [grant number 51378383].

References

Andrade, C., Alonso, C., & Molina, F. (1993). Cover cracking as a function of bar corrosion: Part I-Experimental test. *Materials and Structures, 26,* 453–464.

Arora, P., Popov, B. N., Haran, B., Ramasubramanian, M., Popova, S., & White, R. E. (1997). Corrosion initiation time of steel reinforcement in a chloride environment – A one dimensional solution. *Corrosion Science, 39,* 739–759. doi:10.1016/S0010-938x(96)00163-1

Balafas, I., & Burgoyne, C. J. (2010). Modeling the structural effects of rust in concrete cover. *Journal of Engineering Mechanics, 137,* 175–185.

Bhargava, K., Ghosh, A. K., Mori, Y., & Ramanujam, S. (2005). Modeling of time to corrosion-induced cover cracking in reinforced concrete structures. *Cement and Concrete Research, 35,* 2203–2218. doi:10.1016/j.cemconres.2005.06.007

Caré, S., & Raharinaivo, A. (2007). Influence of impressed current on the initiation of damage in reinforced mortar due to corrosion of embedded steel. *Cement and Concrete Research, 37,* 1598–1612. doi:10.1016/j.cemconres.2007.08.022

Chen, D., & Mahadevan, S. (2008). Chloride-induced reinforcement corrosion and concrete cracking simulation. *Cement & Concrete Composites, 30,* 227–238. doi:10.1016/j.cemconcomp.2006.10.007

Elsener, B. (2005). Corrosion rate of steel in concrete – Measurements beyond the Tafel law. *Corrosion Science, 47,* 3019–3033. doi:10.1016/j.corsci.2005.06.021

Gadve, S., Mukherjee, A., & Malhotra, S. N. (2009). Corrosion of steel reinforcements embedded in FRP wrapped concrete. *Construction and Building Materials, 23,* 153–161. doi:10.1016/j.conbuildmat.2008.01.008

Garboczi, E. J. (2002). Three-dimensional mathematical analysis of particle shape using X-ray tomography and spherical harmonics: Application to aggregates used in concrete. *Cement and Concrete Research, 32,* 1621–1638.

Garcés, P., Andrade, M. C., Saez, A., & Alonso, M. C. (2005). Corrosion of reinforcing steel in neutral and acid solutions simulating the electrolytic environments in the micropores of concrete in the propagation period. *Corrosion Science, 47,* 289–306. doi:10.1016/j.corsci.2004.06.004

González, J., Andrade, C., Alonso, C., & Feliu, S. (1995). Comparison of rates of general corrosion and maximum pitting penetration on concrete embedded steel reinforcement. *Cement and Concrete Research, 25,* 257–264.

Guzmán, S., Gálvez, J. C., & Sancho, J. M. (2012). Modelling of corrosion-induced cover cracking in reinforced concrete by an embedded cohesive crack finite element. *Engineering Fracture Mechanics, 93,* 92–107. doi:10.1016/j.engfracmech.2012.06.010

Häfner, S., Eckardt, S., Luther, T., & Könke, C. (2006). Mesoscale modeling of concrete: Geometry and numerics. *Computers & Structures, 84,* 450–461. doi:10.1016/j.compstruc.2005.10.003

Jaffer, S. J., & Hansson, C. M. (2009). Chloride-induced corrosion products of steel in cracked-concrete subjected to different loading conditions. *Cement and Concrete Research, 39,* 116–125. doi:10.1016/j.cemconres.2008.11.001

Jang, B. S., & Oh, B. H. (2010). Effects of non-uniform corrosion on the cracking and service life of reinforced concrete structures. *Cement and Concrete Research, 40,* 1441–1450. doi:10.1016/j.cemconres.2010.03.018

Lambert, P., Page, C. L., & Vassie, P. R. W. (1991). Investigations of reinforcement corrosion. 2. Electrochemical monitoring of steel in chloride-contaminated concrete. *Materials and Structures, 24,* 351–358. doi:10.1007/Bf02472068

Liu, T., & Weyers, R. W. (1998a). Modeling the dynamic corrosion process in chloride contaminated concrete structures. *Cement and Concrete Research, 28,* 365–379. doi:10.1016/S0008-8846(98)00259-2

Liu, Y., & Weyers, R. E. (1998b). Modeling the time-to-corrosion cracking in chloride contaminated reinforced concrete structures. *ACI Materials Journal, 95,* 675–680.

Manera, M., Vennesland, Øystein, & Bertolini, L. (2008). Chloride threshold for rebar corrosion in concrete with addition of silica fume. *Corrosion Science, 50,* 554–560. doi:10.1016/j.corsci.2007.07.007

Meakin, P., Jøssang, T., & Feder, J. (1993). Simple passivation and depassivation model for pitting corrosion. *Physical Review E, 48,* 2906–2916. Retrieved from http://www.ncbi.nlm.nih.gov/pubmed/9960924

Michel, A., Pease, B. J., Geiker, M. R., Stang, H., & Olesen, J. F. (2011). Monitoring reinforcement corrosion and corrosion-induced cracking using non-destructive X-ray attenuation measurements. *Cement and Concrete Research, 41,* 1085–1094. doi:10.1016/j.cemconres.2011.06.006

Molina, F., Alonso, C., & Andrade, C. (1993). Cover cracking as a function of rebar corrosion: Part 2 – Numerical model. *Materials and Structures, 26,* 532–548.

Moukarzel, C., & Herrmann, H. (1992). A vectorizable random lattice. *Journal of Statistical Physics, 68,* 911–923.

Oh, B. H., & Jang, S. Y. (2005). *Experimental investigation of the threshold chloride concentration for corrosion initiation in reinforced concrete structures.* Paper presented at the 18th International Conference on Structural Mechanics in Reactor Technology, Beijing, China.

Ollivier, J. P., Maso, J. C., & Bourdette, B. (1995). Interfacial transition zone in concrete. *Advanced Cement Based Materials, 2,* 30–38. Retrieved from <Go to ISI>://WOS:A1995QF90100006

Page, C. L., Lambert, P., & Vassie, P. R. W. (1991). Investigations of reinforcement corrosion. 1. The pore electrolyte phase in chloride-contaminated concrete. *Materials and Structures, 24,* 243–252. doi:10.1007/Bf02472078

Pan, Z. C., Chen, A. R., & Ruan, X. (2015). Spatial variability of chloride and its influence on thickness of concrete cover: A two-dimensional mesoscopic numerical research. *Engineering Structures, 95,* 154–169. doi:10.1016/j.engstruct.2015.03.061

Pan, Z. C., Ruan, X., & Chen, A. R. (2014). Chloride diffusivity of concrete: Probabilistic characteristics at meso-scale. *Computers and Concrete, 13,* 187–207. doi: 10.12989/cac.2014.13.2.187

Pan, Z. C., Ruan, X., & Chen, A. R. (2015). A 2-D numerical research on spatial variability of concrete carbonation depth at meso-scale. *Computers and Concrete, 15,* 231–257. doi:10.12989/cac.2015.15.2.231

Poupard, O., Aït-Mokhtar, A., & Dumargue, P. (2004). Corrosion by chlorides in reinforced concrete: Determination of chloride concentration threshold by impedance spectroscopy. *Cement and Concrete Research, 34,* 991–1000. doi:10.1016/j.cemconres.2003.11.009

Pour-Ghaz, M., Isgor, O. B., & Ghods, P. (2009). The effect of temperature on the corrosion of steel in concrete. Part 1: Simulated polarization resistance tests and model development. *Corrosion Science, 51,* 415–425. doi:10.1016/j.corsci.2008.10.034

Poursaee, A., & Hansson, C. M. (2009). Potential pitfalls in assessing chloride-induced corrosion of steel in concrete. *Cement and Concrete Research, 39,* 391–400. doi:10.1016/j.cemconres.2009.01.015

Qian, Z. 2012. *Multiscale Modeling of Fracture Processes in Cementitious Materials* (PhD thesis). Delft University of Technology, Netherlands.

Ruan, X., & Pan, Z. C. (2012). Mesoscopic simulation method of concrete carbonation process. *Structure and Infrastructure Engineering, 8,* 99–110. doi:10.1080/15732479.2011.605370.

Šavija, B., Pacheco, J., & Schlangen, E. (2013). Lattice modeling of chloride diffusion in sound and cracked concrete. *Cement and Concrete Composites, 42,* 30–40. doi:10.1016/j.cemconcomp.2013.05.003.

Schlangen, E. 1993. *Experimental and numerical analysis of fracture processes in concrete* (PhD). Delft University of Technology, Delft.

Schlangen, E., & Garboczi, E. J. (1997). Fracture simulations of concrete using lattice models: Computational aspects. *Engineering Fracture Mechanics, 57,* 319–332. doi:10.1016/S0013-7944(97)00010-6

Schlangen, E., & Van Mier, J. (1992). Simple lattice model for numerical simulation of fracture of concrete materials and structures. *Materials and Structures, 25,* 534–542.

Scrivener, K. L., & Nemati, K. M. (1996). The percolation of pore space in the cement paste/aggregate interfacial zone of concrete. *Cement and Concrete Research, 26,* 35–40.

Shafei, B., Alipour, A., & Shinozuka, M. (2012). Prediction of corrosion initiation in reinforced concrete members subjected to environmental stressors: A finite-element framework. *Cement and Concrete Research, 42*, 365–376. doi:10.1016/j.cemconres.2011.11.001

Wang, Z. M., Kwan, A. K. H., & Chan, H. C. (1999). Mesoscopic study of concrete I: Generation of random aggregate structure and finite element mesh. *Computers & Structures, 70*, 533–544. doi:10.1016/S0045-7949(98)00177-1

Wong, H. S., Zhao, Y. X., Karimi, A. R., Buenfeld, N. R., & Jin, W. L. (2010). On the penetration of corrosion products from reinforcing steel into concrete due to chloride-induced corrosion. *Corrosion Science, 52*, 2469–2480. doi:10.1016/j.corsci.2010.03.025

Yu, H., Shi, X. M., Hartt, W. H., & Lu, B. T. (2010). Laboratory investigation of reinforcement corrosion initiation and chloride threshold content for self-compacting concrete. *Cement and Concrete Research, 40*, 1507–1516. doi:10.1016/j.cemconres.2010.06.004

A site-specific traffic load model for long-span multi-pylon cable-stayed bridges

Xin Ruan, Junyong Zhou, Xuefei Shi and Colin C. Caprani

ABSTRACT

This paper proposes a site-specific traffic load model for long-span multi-pylon cable-stayed bridge. Structural effects are primarily investigated based on influence lines, which are identified as either global effect (GE) or partial effect (PE) depending on the effective influenced region. GEs are further categorised as sensitive effect (SE), insensitive effect (ISE) or less sensitive effect (LSE), considering sensitivity to unbalanced traffic loading. Three on-bridge traffic states are simulated, and Weibull extrapolations are utilised to predict the extreme responses. These responses are analysed and compared with several design codes. Results indicate the maximum response is only 75% of the value calculated based on the design code of China (D60), and even lower than other codes. The responses show strong positive correlation with traffic parameters of annual average daily traffic volume and heavy vehicle proportion, and the on-bridge traffic states have significant influence on the responses. Further, the identified effects of ISE, SE and LSE present different responses, which indicate specific load models are needed accordingly. Finally, a site-specific traffic load model consisting of load form, loading pattern, multi-lane factor and load value is recommended, which gives an accurate illustration on the structural effects and traffic responses.

Introduction

Multi-pylon cable-stayed bridges are a very competitive scheme for large bridges. For these bridges, the traffic load responses, especially unbalanced traffic load action, are always the main focuses (Gimsing & Georgakis, 2011; Virlogeux, 2001). When the single middle span is loaded and the adjacent spans unloaded, the corresponding cables undergo an important tension variation, and the two adjacent pylons are bent towards the load due to the lack of effective constraint of backstays and auxiliary piers, which creates significant bending moments and deflections. To solve this problem, some designers and scholars introduced several structural measures, such as enhancing the flexural rigidity of the middle pylon (Rion–Antirion Bridge in Greece, 268 + 3 × 560 + 268 m); adding cross cables among pylons (Ting Kau Bridge in Hong Kong, 127 + 448 + 475 + 127 m); and setting double bearings between the pylon and girder (Millau Viaduct in France, 204 + 6 × 342 + 204 m), (Bergermann & Schlaich, 1996; Combault & Pecker, 2005; Virlogeux, 1999) etc. These measures are effective. However, as the main span gets longer and the number of pylons increases, it is important to come up with other solutions as well as structural measures. This is particularly important in recent years with several long-span multi-pylon cable-stayed bridges (LSMCBs) constructed such as Jiashao Bridge in China (270 + 5 × 428 + 270 m), Wuhan Erqi Yangtze River Bridge in China (250 + 616 + 616 + 250 m) and Second Forth Bridge in UK (325 + 650 + 650 + 325 m).

Along with the rapid development of sensor and computer technologies, accurate measurement of on-bridge traffic data is now available, and research of traffic load on bridges has received much attention. Precise estimation of traffic load responses on bridges has become a main focus, which may offer breakthroughs for the improvement of cost-effective design and evaluation of LSMCB. Particularly, with the site-specific traffic data, some scholars pay attention to the traffic load on long-span bridges, and facilitate research on the current traffic load models applied to long-span bridges. Buckland (1991) conducted a comprehensive comparison among the codes of the UK, USA and Canada to give effective design guides for bridges with long span, and dynamic or impact load allowance is not recommended as the maximum load occurs with traffic stationary.

Nowak, Lutomirska and Ibrahim (2010) develop the live load models for long-span structures using measured weigh-in-motion (WIM) data, and verify the applicability of the pre-existing HL-93 load model for long-span bridges. Enright, Carey and Caprani (2013) use the micro-simulation method to construct congested flow on long-span bridges, and estimate the suitability of the traffic load models in codes. These studies indicate the current traffic load models given in codes as D60 (Ministry of Communications and Transportation [MOCAT], 2004), American Association of State Highway and Transportation Officials (AASHTO, 2004), BS5400 (BSI, 2006) and Eurocode (EN, 2003) are mostly developed for bridges with small to medium spans. They may not appropriate to be directly used in long-span bridge, and notably, the Eurocode recommends

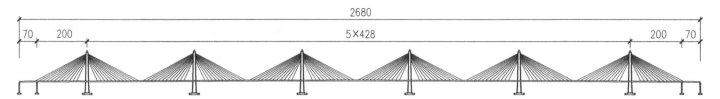

Figure 1. Scheme of a six-pylon cable-stayed bridge (unit: m).

its traffic load model is suitable for the design of bridges with a load length less than 200 m (EN, 2003). In fact, bridge vehicle load responses with long loading length are different from those with small or medium loading length since the respective governing traffic loading scenarios are significantly different (Caprani, 2013; Caprani, O'Brien, & McLachlan, 2008; O'Brien, Hayrapetova, & Walsh, 2012; Ruan, Zhou, & Yin, 2014). Meanwhile, the literature that has particular regulations on the traffic load models for long-span bridges are the ASCE loading (Buckland, 1981) and the superstructure design standard for Honshu–Shikoku Bridge in Japan (Honshu Shikoku Bridge Authority [HSBA], 1980). However, these two models were developed on the base of the traffic data from more than 30 years ago. For more accurate research on LSMCB, a new traffic load model is needed, which should be derived on the basis of more recent traffic data.

WIM is one of the techniques that can record the traffic parameters of vehicle sequence and their weights without disturbing the traffic operation, and so it is widely reviewed and adopted (Hallenbeck & Weinblatt, 2004). Based on WIM data, site-specific traffic load analysis becomes possible (Getachew & O'Brien, 2007; Hajializadeh, Stewart, Enright, & O'Brien, 2015; O'Connor & Eichinger, 2007; Pelphrey et al., 2008). The processes to the assessment of characteristic bridge load responses from site-specific WIM data are similar, and generally can be concluded as follows (Nowak & Rakoczy, 2013; O'Brien, Schmidt, et al., 2015; Ruan, Zhou, & Guo, 2012). Firstly, fit statistical distributions to the WIM data parameters and construct mathematical models of various traffic parameters. Then, simulate random traffic flow under given traffic conditions and parameters. Next, find a suitable distribution to approach the extrema of load responses and extrapolate the extrema to the specified return period. Finally, give the parameter estimation for the distribution of extremes and find the characteristic value for assessment. These procedures provide a new opportunity and solution for the precise modelling of bridge traffic load on long-span multi-pylon cable-stayed bridge (LMCSB).

In this paper, traffic load modelling for LMCSB will be focused on, using the up-to-date site-specific traffic load analysis methodology. At first, to make the load model specific, structural effects in LMCSB are investigated in detail and identified from the perspective of influence lines. Then, random traffic flow is generated by Monte Carlo method on the basis of a 28-day WIM database, and characteristics of load responses are analysed after extrapolation. Finally, a recommended site-specific traffic load model consisting of load form, loading pattern, multi-lane factor and load value is proposed. The research is applied to a six-pylon cable-stayed bridge which has five middle spans with the same length of 428 m. Flat separated steel box girders are used for the

main girder, and its width is 55.6 m with eight lanes of traffic. The pylon is a concrete single column, and the stay-cables are in four spatial planes. Figure 1 gives the scheme of the bridge.

Structural effects with traffic loading

An influence surface/line is a common representative method to account for structural properties in a live load analysis. Hence, influence surfaces of structural effects in LMCSB are investigated in detail here. The influence surfaces are analysed with geometric nonlinearity of dead load included and live load excluded as assumed in current researches (Nowak et al, 2010; Enright et al, 2013).

Global and spatial effects

Multi-pylon cable-stayed bridge is a complex system with various structural effects. For the consideration of traffic loading, some effects present significant response only when the vehicle is applied in a small region compared with the whole loading length available. However, some effects have significant responses despite the vehicle loaded position. Figure 2 gives two influence lines for the considered six-pylon cable-stayed bridge. The most significant influence line for a traffic lane with the maximum values is obtained from the bridge deck influence surface as a simplified demonstration. The influence values are normalised by the maximum absolute value.

It can be observed from the influence line that the vertical girder deflection in the centre of the middle span (VGDMS) changes with the loading region dramatically. Specifically, the positive and negative parts of the influence line alternately appear, and the extreme values and influence areas of positive and negative parts are almost at the same level. That means that when a vehicle is applied anywhere on the whole loading length a significant response can be created. Hence, the structural effects that have the same features of influence lines like VGDMS can be regarded as global effects (GEs), which are highly related to the global behaviour of traffic loading.

Conversely, for the influence line of longitudinal girder bending moment in the centre of the middle pylon (LGBMS), it can be noticed that the effective loading region is only around the centre of the middle span since the influence values in other regions are negligible. Further, the extreme values and influence areas of positive and negative parts in LGBMS significantly vary. This means that considerable response can only be created when a vehicle is applied in the effective influence region which is far less than the loading length. Therefore, this type of load effect can be treated as a partial effect (PE), which is related to limited short range behaviour of traffic loading.

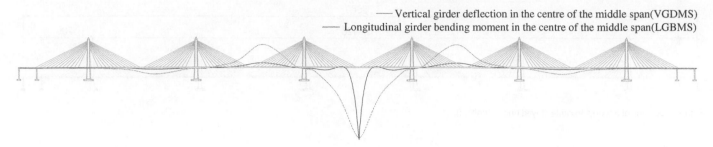

Figure 2. Comparison between the influence lines of global and PE.

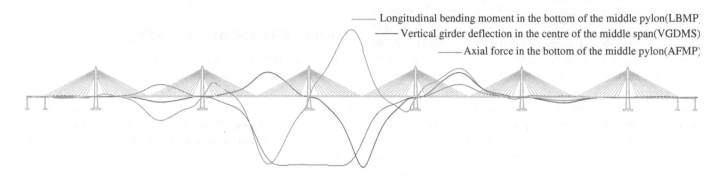

Figure 3. Comparison of influence lines for effects with various sensitivities to unbalanced traffic loading.

Sensitivity to unbalanced traffic loading

GEs have long effective loading length, and the positive and negative parts of the influence lines are at the same level, which mean that the features of GEs can be diverse when unbalanced traffic loading is considered. For example, some of the GEs may have equivalent parts of positive and negative influence areas, and then unbalanced traffic loading will create the most adverse response. However, some of the GEs may have positive (or negative) part of influence areas far larger than the other, and consequently balanced traffic loading will contribute to the maximum responses. Figure 3 shows three influence lines of the six-pylon cable-stayed bridge. As before, the influence values are normalised by the maximum absolute value.

As depicted in the figure, the influence line of longitudinal bending moment in the bottom of the middle pylon (LBMP) presents the same level of extreme influence values and influence areas in positive and negative parts along with the loading region. This means that the arrangements of unbalanced traffic loading will lead to the most unfavourable responses for such influence lines. Hence, load response of structural effects like LBMP will be sensitive to unbalanced traffic loading, and they are defined as sensitive effects (SEs). Alternatively, the influence line of the axial force in the bottom of the middle pylon (AFMP) demonstrates another situation. The influence origins of AFMP almost keep the same negative sign along the loading region, and the extreme influence values and influence areas in positive and negative parts are dramatically different. Thus, arrangements of balanced traffic loading will create the most unfavourable responses on structural effects like AFMP, and these effects can be treated as insensitive effects (ISEs). Finally, structural effects like vertical girder deflection in the centre of the middle girder (VGDMS), whose extreme influence values and influence areas in positive

and negative parts are of similar magnitudes, can be identified as a less sensitive effect (LSE).

To summarise, load responses of ISEs are mainly determined by the average vehicles weights in the loading region. Load responses of SEs are highly related to vehicular arrangement as well as average vehicle weights. Lastly, for LSEs, features of both ISEs and SEs are coupled. These features of GEs in LMCSB are essential for the construction of a site-specific traffic load model.

Metrics of structural effects

With the GE, PE, SE, ISE and LSE described above, the extreme influence values, influence areas in positive and negative parts, and effective influence region are critical to identify the relevant structural effects. Hence, metrics of the influence line are proposed as follows:

$$y_{ma} = max(|y(x)|) \cdot$$

$$L_{eff} = \int_0^L w dx; w = \begin{cases} 0, & y(x) \in [-0.2y_{ma}, 0.2y_{ma}] \\ 1, & y(x) \in [0.2y_{ma}, y_{ma}] \cup [-y_{ma}, -0.2y_{ma}] \end{cases}$$

$$R_{Are} = min(|R_{Are+}/R_{Are-}|, |R_{Are-}/R_{Are+}|),$$

$$R_{Are-} = \int_0^L y(x)dx, y(x) \leq 0,$$

$$R_{Are+} = \int_0^L y(x)dx, y(x) \geq 0;$$

$$R_{Ext} = min(|R_{Ext+}/R_{Ext-}|, |R_{Ext-}/R_{Ext+}|),$$
$$R_{Ext-} = min(y(x)), R_{Ext+} = max(y(x)) \tag{1}$$

where L is the loading length of the structural effect; x is the position of the differential length; $y(x)$ is the influence function of the structural effect, representative of the influence surface, y_{ma} is the maximum absolute influence value; L_{eff} is the effective

Table 1. Metrics for classification of structure effects.

Structural effect		R_{Are}	R_{Ext}	L_{eff}
Global effect	Sensitive	.8–1.0	.8–1.0	.1–1.0L
	Less sensitive	.2–.8	.2–.8	
	Insensitive	.0–.2	.0–.2	
Partial effect	/	/	/	.0–.1L

influence length of the structural effect; R_{Are+} is the accumulated positive influence area of the structural effect; R_{Are-} is the accumulated negative influence area of the structural effect; R_{Are} is the minimum absolute ratio of the influence areas between positive and negative parts; R_{Ext+} is the maximum extreme influence value in the positive parts; R_{Ext-} is the minimum extreme influence value in the negative parts; R_{Ext} is the minimum absolute ratio of the extreme influence values between positive and negative parts.

In these metrics, GEs and PEs are identified mainly by L_{eff} which is the accumulated length where the influence value is larger (positive) or smaller (negative) than the 20% of the maximum absolute influence value. If L_{eff} is less than or equal to 10% of the total loading length L ($L_{eff} \leq 0.1L$), the effect is regarded as a PE, and a GE if more than 10%, respectively. As for the GEs, if R_{Ext} or R_{Are} is within the range of .8–1.0, the effect can be recognised as a SE considering unbalanced traffic loading. Accordingly, if R_{Ext} and R_{Are} are in the range of .0–.2, the effect can be identified as an ISE, and a LSE is in the other conditions. Identified values of the metrics are illustrated in Table 1.

Structural effects in LMCSB

It is obvious that there are many components in LMCSB, and structural effects of different components are relatively complex. Therefore, it is important to pick out representative structural effects for analysis purpose. Table 2 provides the representative structural effects of the six-pylon cable-stayed bridge in general consideration. Using the proposed metrics, the effects can be systematically categorised. Metrics values of influence lines for the representative structural effects are listed, and the corresponding locations of the effects are provided in Figure 4. It is noticed that the longitudinal girder bending moment is a typical PE, and the longitudinal pylon bending moment is a SE.

PEs in cable-stayed bridge are more like the structural effects in short or medium span bridges due to their limited effective influence length, and so their characteristics of governing traffic loading are different from GEs. Hence, in this paper, only GEs are further studied. It is taken that the GEs in the middle span are of more concern than the other spans, therefore five representative GEs, including the axial force in the bottom of the middle pylon (AFMP), the longitudinal bending moment in the bottom of the middle pylon (LBMP), the vertical girder deflection in the centre of the middle span (VGDMP), the axial force in the girder adjacent to the middle pylon (AFGP) and the axial force of the centre cable in the middle span (AFCC) are examined using site-specific traffic load analysis.

Table 2. Classification metrics of structure effects in a six-pylon cable-stayed bridge.

Representative effects	Location	Metrics of influence line			Category
		R_{Are}	R_{Ext}	L_{eff}	
Longitudinal girder bending moment	Centre of the third middle span/A	/	/	.0254	PE
	Centre of the second middle span/B	/	/	.0261	PE
	Centre of the middle span/C	/	/	.0257	PE
Vertical girder deflection	Centre of the third middle span/A	.4886	.3439	.2228	LSE
	Centre of the second middle span/B	.6706	.3374	.2690	LSE
	Centre of the middle span/C	.6629	.3476	.2780	LSE
Longitudinal pylon bending moment	Bottom of the side pylon/H	.3763	.5213	.2813	LSE
	Bottom of the second side pylon/I	.8636	.9873	.3918	SE
	Bottom of the middle pylon/J	.9741	.9972	.4388	SE
Axial pylon force	Bottom of the side pylon/H	.1255	.2141	.2604	ISE
	Bottom of the second side pylon/I	.0930	.1951	.2925	ISE
	Bottom of the middle pylon/J	.1291	.1901	.3593	ISE
Axial girder force	Adjacent to the side pylon/D	.8773	.5768	.2713	ISE
	Adjacent to the second side pylon/E	.5949	.3194	.3093	LSE
	Adjacent to the middle pylon/F	.5676	.2901	.3750	LSE
Axial cable force	Centre cable in the middle span/M	.4220	.2672	.2634	LSE
	Quartile cable in the middle span/N	.5784	.4746	.3515	LSE

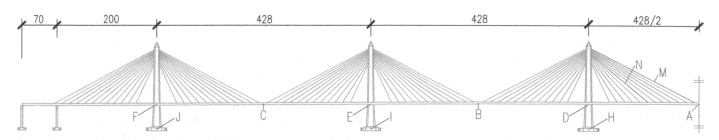

Figure 4. Locations of the featured effects.

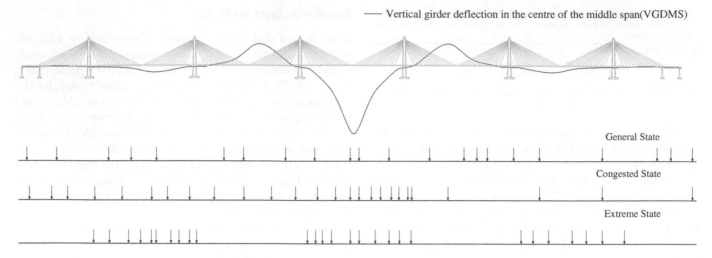

Figure 5. Loading scenarios of various on-bridge traffic states for a structural effect.

Loading responses under random traffic flow

General processes for the assessment of characteristic bridge load responses from site-specific WIM data are given in the introduction. In the study of bridges with long loading length, the traffic running states of random traffic simulation and extreme extrapolation are critical issues (Barceló, 2010; Enright et al., 2013; Nowak et al., 2010; O'Brien et al., 2012), which will be emphasised in the following.

On-bridge traffic states

Due to the complexity of the influence lines in LMCSB, the critical traffic loading scenarios for various structural effects can be quite different. Free-flow, congested-flow and blocking-flow are three on-road traffic running states generally considered (Barceló, 2010). Similarly, three types of traffic loading scenarios are considered in long loading length as general (free-flow) state, congested state (CS) and extreme state, respectively.

As depicted in Figure 5, the vehicle flows for the three on-bridge traffic states are represented with arrows indicating the gross vehicle weights, and the influence line of the vertical girder deflection in the centre of the middle span (VGDMS) is given as an example. It is indicated that higher vertical deflection as well as lower occurrence probability of vehicles arrangement are expected under the CS than the general state (GS). With the locally arranged vehicles under the extreme state, the worst scenario occurs when all the vehicles are loaded in the regions with the same sign of influence values, in which case the results can be extremely unfavourable and the probability of occurrence is correspondingly quite low.

Generally, these three on-bridge traffic states represent various features of responses and occurrence probabilities. However, they all contribute to the extreme traffic load responses of LMCSB, and thus all the three on-bridge traffic states will be included in the site-specific traffic load analysis.

Monte Carlo simulation

WIM data give a comprehensive record of the traffic flow but due to the running costs and durability of WIM systems, long-run traffic data are needed to account for the extreme load response that may appear in the required return period. Monte Carlo is a common simulation approach that uses statistical distributions derived from the WIM data and generates long periods of traffic, including vehicles and combinations of vehicles that have not been observed during the period of measurement (Enright & O'Brien, 2013).

In this study duration of 28 consecutive days WIM data are utilised, and the general vehicle features such as weight, axle weight, speed, configuration, traffic volume and gap, etc. are fitted and modelled (Ruan & Zhou, 2014). In the simulation of traffic flows for long-span bridges, the inter-vehicle gap and the traffic accidents models are important (O'Brien, Lipari, & Caprani, 2015; Wu et al., 2015). Generally, the Poisson distribution is utilised for the vehicle arrival model in free-flow traffic and the binomial distribution is adopted for the congested-flow traffic (Barceló, 2010; Ruan et al., 2012). The traffic accident model is developed by simulating traffic congestions, and the locations of the resulting traffic jams are generally considered in the centre of span or near to pylon. The probability of accident occurrence is adopted according to the data collected from highways in China (Ruan et al., 2012). With these statistical models, a one-year random vehicle sequence of free-flow and congested flow are generated by the MC method.

Extreme extrapolation

The D60 (MOCAT, 2004) specifies a characteristic traffic load response as that value which has a 5% probability of being exceeded in 100 years, which is equivalent to the value with a return period of approximately 1950 years. Extrapolation to extremes from the simulation runs that cover a short duration is the critical method that produces an estimation of the characteristic value, but may have relatively high variability.

Various extrapolation methods are reviewed and researched, which are known as normal probability extrapolation, rice formula extrapolation and extreme value theory-based prediction, etc. Extreme value theory-based prediction gives a relaxed requirement on the sample data, and conducts a relatively precise and stable approach using the tail of the traffic responses

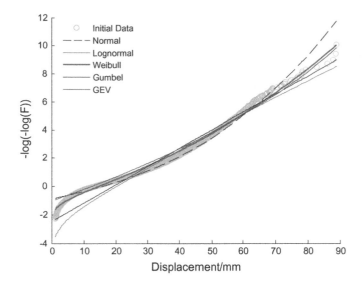

Figure 6. Load responses fitting for VGDMS.

when the data duration is long (O'Brien et al., 2015). Since the simulation is one year, the extreme value theory-based prediction is utilised.

In Extreme Value Theory, a parent distribution is assumed to model the observed data, and an asymptotic distribution is introduced to approximate the maximum or minimum value of the parent distribution for predication (Castillo, Hadi, Balakrishnan, & Sarabia, 2004). If the parent distribution is precise enough to describe the observed data, and is also belongs to maximum attraction domain, then the canonical transformed asymptotic distribution converges to three types of extreme value distributions, Gumbel, Frechet and Weibull distribution, individually. These three types can be combined into one distribution, which is known as generalised extreme value (GEV) distribution:

$$H(x; \xi, \sigma, \mu) = exp\left\{-(1+\xi\frac{x-\mu}{\sigma})^{-1/\xi}\right\}$$
$$where\ 1+\xi(x-\mu)/\sigma < 0$$

(2)

where μ, σ, α are the mean value, standard value and shape parameter of the observed data, respectively. The type II and type III classes of extreme value distribution ξ correspond, respectively, to the cases $\xi > 0$ and $\xi < 0$ in this parameterisation. The subset of the GEV family with $\xi = 0$ is interpreted as the limit of Equation (4) as $\xi \to 0$, leading to the Gumbel family.

Assuming that the yearly traffic data are identically distributed (i.e. no traffic growth), and the traffic load responses of the one-year simulation runs can represent the data characteristic of other years. The GEV distribution and GEV family of Gumbel and Weibull distribution are applied to fit the one-year simulated traffic load responses with attention to the tail of the cumulative distribution function, and the results are extrapolated to the return period of 1950 years. In order to demonstrate the tail tendency and goodness of fit of the parent distribution, Gumbel probability paper is used and traditional empirical distributions such as Normal and Lognormal are also compared with the parameters estimation of maximum likelihood method. Figure 6 gives an example of load response fitting for a representative effect of VGDMG. It is showed Normal and Lognormal distributions are relatively poor in fitting the data, while the Weibull

gives a better approach on the tail tendency than Gumbel and GEV. Other representative effects almost show the same results, and therefore the Weibull extrapolation is used to assess the characteristic traffic load response with 95% assurance rate in a 100-year design reference period as illustrated below:

$$F_{100}(x) = Pr\{X_1 \leq x\} \times Pr\{X_2 \leq x\} \times \cdots \times Pr\{X_{100} \leq x\}$$
$$= F(x)^{100}$$
$$x = \mu + \sigma\left[-\ln(1 - 0.95^{0.01})\right]^{-\xi}$$

(3)

where $F_{100}(x)$ is the distribution of load responses in 100 years and $F(x)$ is the Weibull distribution of yearly load responses, μ, σ, α are the estimated mean value, standard value and shape parameter of the Weibull fitting on the yearly load responses with maximum likelihood method.

Characteristic of load responses

With the representative structural effects and simulated random traffic flow, the load responses for one year of the three on-bridge traffic states can be acquired. In the GS, traffic conditions with free-flow are arranged on the whole influence surface, while in the CS, traffic conditions with congested-flow are applied. Finally, in the extreme state, traffic conditions with congested flow are involved and only arranged in the influenced regions with the same sign of influence value. To compare with the results calculated by the traffic load model in D60, the load responses are all extrapolated to the characteristic value with 95% assurance rate in a 100-year design reference period.

Two traffic parameters of annual average daily traffic volume (AADT) and heavy vehicle proportion (HV) are studied to explore the load response under different traffic parameters. From the statistical analysis of the 28-day WIM data, the AADT of 8-lane traffic is close to 38,000 and the HV approaches 12%. The traffic volume in the highway is far from saturation and the heavy vehicle proportion represents the general freightage in China. Therefore, the values of the considered AADT are 40,000, 60,000, 80,000, 100,000 and 120,000, and the HV are 5, 10 and 20%, respectively. Heavy vehicles refer to vehicles with gross weight more than 20t. These two parameters are essential factors on the structural effect with long loading length (Buckland, 1991; Nowak & Rakoczy, 2013). The comparisons between actual responses of various effect types in three on-bridge traffic states are presented in Figure 7, and standard value calculated by the traffic load model of D60 are also given. The abscissa represents standardised AADT of the four-lane bidirectional traffic and the ordinates represent the ratio of actual to standard value.

Figure 7 indicates that the responses of effects are positively correlated with AADT and HV. Particularly, as for the GS and CS with AADT lower than 80,000, the responses are almost linear with AADT and HV. When the AADT is higher than 80,000, the randomness of the vehicles and traffic flows are less apparent, which leads to a lesser correlation between the structural effects and the AADT and HV. In this case, the increments of the effects except AFMP slow down with the increasing AADT.

It is also noticed that the ratios of actual structural responses and standard values are almost within the range of 20–40% in

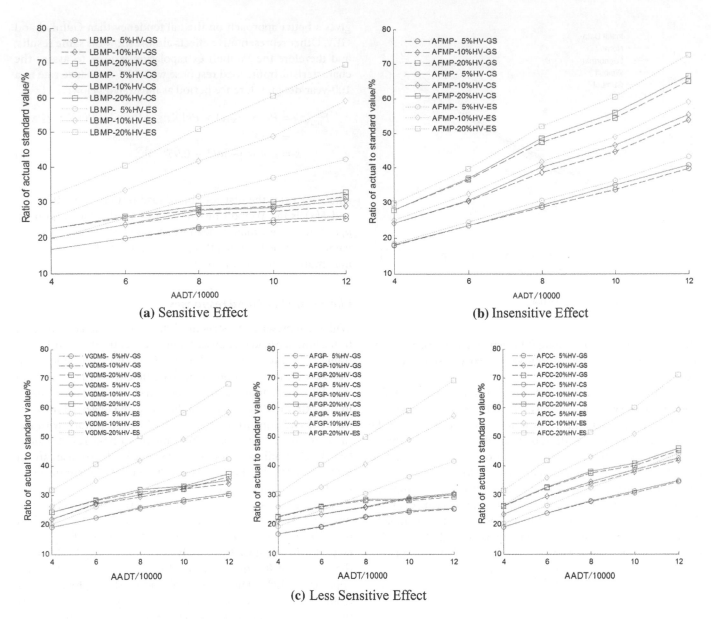

Figure 7. Load responses under three traffic running states and code model for a six-pylon cable-stayed bridge.

GS except AFMP, 30–50% in CS, and up to 70% in extreme state (ES). The results indicate that traffic states significantly affect the extreme values. However, even in the most unfavourable extreme state, the actual responses are still much lower than the code values, demonstrating that the current D60 is conservative for LMCSB.

Comparison among the GEs indicates different effect types show various levels of sensitivity to the on-bridge traffic states. For the SE as LBMP, the traffic state is the most sensitive factor, as expected. For the ISE as AFMP, the results are not much influenced by the traffic states but quite dependent on the AADT and HV. As for the LSEs as VGDMS, AFGP and AFCC, their responses are dependent on the traffic states and linear with the AADT and HV as well.

In conclusion, the actual traffic load responses are lower than standard value, even if the load factors of 1.4 for live load in D60 is included, there is still a large scope for rationalisation of the traffic load model in D60 for LMCSB. Responses of effects with different types in LMCSB are strongly correlated and vary significantly with the traffic parameters and on-bridge traffic states, which prove it is essential to make a distinction between SE, LSE and ISE in a site-specific traffic load model, and further, the influences of traffic parameters and on-bridge traffic states should be included.

Current traffic load models from codes

Several codes including D60, BS5400, AASHTO, ASCE loading, HSBA and Eurocode are analysed to compare with the site-specific load responses. Parameters from different codes are listed in Table 3. In D60, lane load model is adopted for long-span bridge analysis, and in AASHTO both Truck Model 1 and lane load model are used. Load Combination 1 with HA plus HB in BS5400 are applied, and Load Model 1 with tandem system and uniformly distributed load (UDL) in Eurocode, the equivalent L load in Load Model 2 of HSBA, and uniform load and concentrated load with loading length over 1950 m and 100% HV in ASCE loading are utilised, respectively.

Table 3. Traffic load models and their parameters from different codes (for loaded length of 2680 m and eight-lane traffic).

Codes	D60	BS5400	AASHTO	ASCE loading	HSBA	Eurocode
Uniform lane load	10.5 kN/m	9 kN/m	9.3 kN/m	10.5 kN/m	2.12 kN/m^2(5.5 m)+1.06 kN/m^2	2.5/9 kN/m^2
Concentrated lane load	360 kN	120 kN	35 + 145+145 kN	747 kN	128(5.5 m) + 2.5 kN/m	200/400/600 kN
Multi-lane factor	.5	/	.65	.51	/	/
Longitudinal reduction factor	.93	/	/	/	/	/
Dynamic impact factor	.05	/	/	/	/	/

Table 4. Comparison of the load responses from actual traffic flows and the codes for the representative effects of a six-pylon cable-stayed bridge with 100,000 AADT and 20% HV.

Traffic load		VGDMS ($\times 10^2$)/mm	AFGP ($\times 10^6$)/kN	AFCC ($\times 10^6$)/kN	LBMP ($\times 10^8$)/kN m	AFMP ($\times 10^7$) /kN
Actual traffic flow	General state	.81	1.17	.44	1.38	1.50
	Congested state	.83	1.19	.45	1.43	1.54
	Extreme state	1.29	2.69	.70	3.02	1.68
Code models	D60	2.23	3.87	.98	4.35	2.31
	BS5400	2.03	3.90	1.01	3.92	2.34
	AASHTO	2.33	4.09	1.03	4.59	2.49
	ASCE loading	2.51	4.25	1.09	4.79	2.41
	HSBA	2.07	3.79	.94	3.75	2.13
	Eurocode	5.69	9.91	2.57	11.10	6.19

For eight-lane loading conditions, multi-lane factor given by D60, AASHTO and ASCE loading are .5, .65 and .51, respectively. In BS5400, Eurocode and HSBA, the load value applied in lateral lanes are different, which is similar to the multi-lane factor. With regard to the bridge loading length, these codes have certain regulations considering their applications. For Eurocode, the load model is only applicable to bridges with spans less than 200 m. For AASHTO, bridge spans were expected to be less than 500 feet (152.4 m) until the 7th (2014) edition in which the HS20 load model is adopted for long-span bridges. A longitudinal reduction factor of .93 is adopted for bridges longer than 1000 m in JTG D60-2004. The HA UDL in BS5400, uniform load in ASCE loading and equivalent L load in HSBA reduce with the increasing length, which can be considered as longitudinal reduction factors. In BS5400, AASHTO, Eurocode, ASCE loading and HSBA, dynamic impact is included in the load model, while, in D60, the dynamic factor is excluded, therefore, a factor of .05 recommended for long-span bridge is additionally used for the equal comparisons.

Table 4 presents the load responses induced by the actual traffic flows and the codes. The responses of the actual traffic flows are calculated with 100,000 AADT and 20% HV, which is a reasonable traffic state for the eight-lane highway in China. The responses of the codes are obtained by arranging the load models on the influence surface with the same sign of influence value. It is concluded that the regulations of effect types toward the codes vary, and the structural responses derived from the actual traffic flows are all lower than those from the codes. The Eurocode gives the highest values while others are generally on the same level, which can be explained by the fact that the applicability of the load model in Eurocode is restrained to the bridges with spans less than 200 m. It is important to note that the codes are typically calibrated using both load and resistance factors to achieve a target level of safety (i.e. structural reliability), and comparing only loadings does not give a true indication of the level of safety of any particular code.

It is apparent that for the bridge in consideration, the current traffic load models in the national codes provide loading that is far more onerous than the traffic data studied. Therefore, it is necessary to improve the traffic load model to satisfy the requirements for LMCSB.

Site-specific traffic load model

A traffic load model is a simplified tool for the bridge design and evaluation. A site-specific traffic load model should account for the extreme load responses to which the bridge in the referenced period will be subjected with acceptable reliability. On the other hand, the model should also be as simple and understandable as possible for engineering applications. Load form, loading pattern, multi-lane factors and load values are four essential parts of a comprehensive traffic load model, which are presented as follows.

Load form

Load forms of the current codes are reviewed in Table 5. Generally, there are three types for global design of bridges, namely uniform line load with concentrated load, uniform line load and uniform surface load with concentrated load, respectively. Concentrated load generally represents a heavy truck which can cause considerable responses when loaded in the region with large influence values. Uniform load indicates that the average load level is crucial for structural effects, which is widely recommended in design for long loading lengths. Uniform line loads seem to be more reasonable than a uniform surface load due to the inherent interaction between the vehicle flow and bridge deck.

GEs in LMCSB have the features of long effective loading length, hence, a uniform line load form is recommended. With consideration of structural effects in pylons that are indirectly connected with bridge deck, the influence surfaces are generally flat and smooth, and thus a concentrated load form is not recommended. However, for the components of cable, girder and pier that are directly connected to the bridge roadway, the influence surfaces of these effects present sharp peaks in some locations, and a concentrated load form is suggested. Specifically, a uniform line load form is suggested for AFMP and LBMP, and a uniform

Table 5. Load forms in design codes.

Design codes	Load form	
	Overall design	Partial design
D60	Uniform line load with concentrated load	Single truck
BS5400	Uniform line load with concentrated load	Single truck
ASSHTO	Uniform line load or uniform line load with concentrated loads	Single truck or uniform line load with concentrated load
ASCE loading	Uniform line load with concentrated load	/
HSBA	Uniform surface load with uniform line load	Single truck
Eurocode	Uniform surface load with concentrated loads	Single truck

Table 6. Traffic loading patterns for LMCSB (see Figure 8 for definition of terms).

Structural effects		On-bridge traffic states		
		General state	Congested state	Extreme state
Girder	VGDMS	$R_c + R_d$		Arranged on the influence region with the same sign of influence value
	AFGP	$R_b + R_c$ or $R_d + R_e$		
Pylon	AFMP	$R_a + R_b + R_c + R_d$ or $R_c + R_d + R_e + R_f$		
	LBMP	$R_a + R_b$ or $R_c + R_d$		
Stay cable	AFCC	$R_d + R_b$ or $R_c + R_e$		

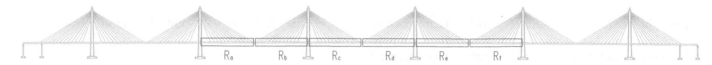

R_a R_b R_c R_d R_e R_f

Figure 8. Divided spans of traffic loading pattern for a six-pylon cable-stayed bridge.

line load with concentrated load form is suggested for VGDMS, AFGP and AFCC accordingly.

Loading pattern

In the current traffic load models, the load forms are expected to be arranged on the influence regions with the same sign of influence value, which can create the most unfavourable load response. However, this arrangement usually gives a very low probability of occurrence in actual conditions, and it should only be considered for the ultimate condition.

A list of loading patterns for the effects is suggested in Table 6 for the three on-bridge traffic states. In the extreme state, the

ultimate structural performances are concerned, and the loading pattern is recommended to be the same as the current codes. In general and CSs, structural serviceability or other performances may be the focus. Therefore, each span is divided into equal halves, and the loading regions only involve two adjacent spans. Figure 8 gives the divided spans for the six-pylon cable-stayed bridge. Taking VGDMS as an example, in general and CSs, only vehicles in the regions of R_a to R_f can create considerable responses. The girder sags when the load is arranged in regions of R_c and R_d, and hogs with the load arranged in regions of R_a, R_b, R_e and R_f. When the uniform line load is loaded in the regions of R_c and R_d, the concentrated load should be added at the position with the extreme influence value within the regions of R_c and R_d. Loading patterns for other effects can be obtained in a similar way.

Multi-lane factor

A multi-lane factor is generally utilised in the analysis of bridges with multiple lanes since a traffic load model is generally abstracted from the lane with maximum extreme load responses. In D60, the multi-lane factor are derived based on the assumptions that the traffic loads are independently and identically distributed on different lanes (Bakht & Jaeger, 1990). However, it is known that load responses in the lateral lanes are quite different (Hallenbeck & Weinblatt, 2004; Ruan & Zhou, 2014).

Load responses of bending moments in the middle girder of simple supported bridge with length varying from 5 to 1000 m are investigated to present the response correlations among lateral lanes. Characteristic values of 98% acceptability in each lane (*R-lane-0.98*) are depicted in Figure 9. The responses are normalised by the total (sum of all lanes) characteristic responses (*R-total-0.98*) with the same acceptability. It can be concluded

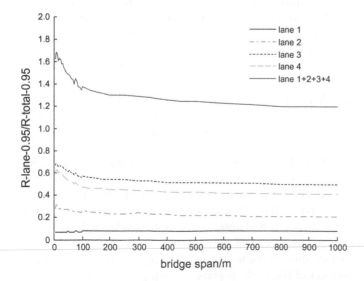

Figure 9. Characteristic traffic load responses of bending moment in middle span of simple supported bridge with length varied from 5 to 1000 m.

Table 7. Load values for structural effects in LMCSB.

Structural effects		Insensitive effect	Less sensitive effect	Sensitive effect
		AFMP	VGDMS, AFGP, AFCC	LBMP
Standard value q_k		3.40	3.10	2.90
On-bridge traffic states/η_1	General state	1.00	1.00	1.00
	Congested state	1.01	1.05	1.05
	Extreme state	1.02	1.12	1.20
AADT/η_2	40,000–120,000	.60–1.40	.80–1.20 (general and congested state)	.60–1.40 (extreme state)
HV/η_3	5%		1.00	
	10%		1.40	
	20%		1.70	

that when the loading length is lower than 200 m the normalised values are not stable and present high variability. However, when the loading length is higher than 200 m, the relative ratio becomes comparatively stable and converges to a constant value. Which means the relative response ratio between lateral lanes becomes constant, and the extreme load responses of different lanes in the same direction are not the same value. Further, the characteristic responses of lane 1 plus lane 2, lane 3 and lane 4 are nearly 1.22 times than the characteristic total responses, which indicate the extreme lane load response will not appear simultaneously, a coefficient of 1/1.22 = .82 is added if the lane load model is treated as the same.

According to the Figure 9, when the traffic load model is constructed based on the maximum lane load response e.g. lane 3, the modified factors for lane 1, lane 2 and lane 4 are .2, .5 and .9 respectively. Therefore, the multi-lane factor for long-span bridge is $(.2 + .5 + 1.0 + .9) \times .82/4 = .533$. In the following, the equivalent multi-lane factor of .533 will be used in the calibration of load value of traffic load model.

Load value

Load values of the site-specific traffic load model are calibrated on the basis of the structural responses in Figure 7 and the aforementioned load forms, loading patterns and multi-lane factor. Since the load responses are highly correlated with traffic parameters and on-bridge traffic states, and the responses of effect types varied, a general expression of the load value is given by:

$$Q_k = \eta_1 \eta_2 \eta_3 q_k \qquad (4)$$

where Q_k is the characteristic value of the traffic load response; q_k is the standard value of the traffic load response, which represents the traffic condition with 80,000 AADT and 5% HV in GS; η_1 is the correction coefficient for on-bridge traffic states; η_2 is the correction coefficient for traffic parameter of AADT; η_3 is the correction coefficient for traffic parameter of HV.

Load values for sensitive, less sensitive and insensitive structural effects are presented in Table 7. The concentrated load for the effects in cable and girder is unified as a constant value, and it is concluded from the WIM data that 95% of the trucks weigh less than 40t (Ruan & Zhou, 2014). Therefore, a concentrated load of 400 kN is recommended. The standard value for ISEs is the highest, and that of the SEs is the lowest. The characteristic values can be adjusted from the standard values according to coefficients of on-bridge traffic states, AADT and

HV. Linear interpolation is used when the target values of traffic parameters falls between the given ones. As for coefficients for AADT in SE and LSE, the coefficients of .6–1.4 are suggested in extreme state, while for other states, coefficients of .8–1.2 are suggested.

In the most unfavourable condition with 120,000 AADT and 20% HV in extreme state, the load values for ISE, LSE and SE are almost in the same level as $3.4 \times 1.02 \times 1.4 \times 1.7 = 8.25$, $3.1 \times 1.12 \times 1.4 \times 1.7 = 8.26$ and $2.9 \times 1.2 \times 1.4 \times 1.7 = 8.28$, respectively. With the consideration of multi-lane factor of .533, the uniform line load in the proposed traffic load model under the extreme state is $8.28 \times .533 \times 4 \times 2 = 35.3$ kN/m. In contrast, the uniform line load in D60 is $10.5 \times .50 \times 8 = 42.0$ kN/m, which shows that the proposed model is about 15.95% lower than D60, fitting well with the results of Figure 7. The comparison between these two values shows that the proposed site-specific traffic load model is rational for LMCSB.

Conclusions

Based on the presented study, the following results can be derived:

(1) Structural effects of LMCSB are numerous and complicated. The new definitions of GE, PE, SE, ISE and LSE are essential to make the load model specific, and the proposed metrics from the influence line are reliable for the identification of structural effects.

(2) The random traffic load responses of the six-pylon cable-stayed bridge indicate that with the same reliability requirement, the maximal responses are about 75% of the standard value calculated by the design codes of China (D60), and even lower than other codes. The load responses show strong positive correlation with traffic parameters of annual average daily traffic volume (AADT) and heavy vehicle proportion (HV), and the on-bridge traffic states have significant influence on the responses. Furthermore, the identified effects of ISE, SE and LSE present different responses, which indicate that specific load models are needed accordingly.

(3) A site-specific traffic load models for LMCSB is proposed. Load form, loading pattern, multi-lane factor and load value are adjusted and calibrated. The proposed load model is about 15.95% smaller than the corresponding value in D60 under the extreme state, which gives an accurate illustration on the features of structural effects and traffic responses.

Nomenclature

LMCSB   long-span multi-pylon cable-stayed bridge;

SE   sensitive effect;

LSE   less sensitive effect;

AADT   annual average daily traffic volume;

GEV   generalised extreme value;

AFMP   axial force in the bottom of the middle pylon;

LBMP   longitudinal bending moment in the bottom of the middle pylon;

VGDMS   vertical girder deflection in the centre of the middle span;

AFGP   axial force in the girder adjacent to the middle pylon;

AFCC   axial force of the centre cable in the middle span;

GE   global effect;

PE   partial effect;

ISE   insensitive effect;

HV   heavy vehicle proportion;

UDL   uniformly distributed load.

Disclosure statement

No potential conflict of interest was reported by the authors.

Funding

This work was supported by National Nature Science Foundation of China [grant number 51108338], [grant number 51478337]; Fundamental Research Funds for the Central Universities in China.

References

American Association of State Highway and Transportation Officials. (2004). *AASHTO LRFD Bridge design specifications* (3rd ed.). Washington, DC: Author.

Bakht, B., & Jaeger, L. G. (1990). Bridge evaluation for multipresence of vehicles. *Journal of Structural Engineering, 116*, 603–618.

Barceló, J. (2010). *Fundamentals of traffic simulation.* Heidelberg: Springer.

Bergermann, R., & Schlaich, M. (1996). Ting Kau Bridge, Hong Kong. *Structural Engineering International, 6*, 152–154.

BS 5400-2. (2006). *Specification for loads.* London: British Standards Institution.

Buckland, P. G. (1981). Recommended design loads for bridges. *ASCE Journal of the Structural Division, 107*, 1161–1213.

Buckland, P. (1991). North American and British long-span bridge loads. *Journal of Structural Engineering, 117*, 2972–2987.

Caprani, C. (2013). Lifetime highway bridge traffic load effect from a combination of traffic states allowing for dynamic amplification. *Journal of Bridge Engineering, 18*, 901–909.

Caprani, C. C., O'Brien, E. J., & McLachlan, G. J. (2008). Characteristic traffic load effects from a mixture of loading events on short to medium span bridges. *Structural Safety, 30*, 394–404.

Combault, J., & Pecker, A. (2005). Rion-Antirion bridge, Greece-concept, design and construction. *Structural Engineering International, 15*(1), 22–27.

Enright, B., Carey, C., & Caprani, C. (2013). Microsimulation evaluation of Eurocode load model for American long-span bridges. *ASCE Journal of Bridge Engineering, 18*, Special Issue: Eurocodes and their implications for bridge design: Background, implementation, and comparison to North American practice, 1252–1260.

Enright, B., & O'Brien, E. J. (2013). Monte Carlo simulation of extreme traffic loading on short and medium span bridges. *Structure and Infrastructure Engineering, 9*, 1267–1282.

Eurocode1 Part2. (2003). *Traffic loads on bridges.* Washington, D.C: European Committee for Standardization.

Castillo, E., Hadi, A. S., Balakrishnan, N., & Sarabia, J. M. (2004). *Extreme value and related models with applications in engineering and science.* Hoboken, NJ: Wiley-Interscience.

Getachew, A., & O'Brien, E. J. (2007). Simplified site-specific traffic load models for bridge assessment. *Structure and Infrastructure Engineering., 3*, 303–311.

Gimsing, N. J., & Georgakis, C. T. (2011). *Cable supported bridges: Concept and design.* New Jersey: Wiley.

Hallenbeck, M. E., & Weinblatt, H. (2004). *Equipment for collecting traffic load data.* Washington, DC: Transportation Research Board (No. 509).

Hajializadeh, D., Stewart, M. G., Enright, B., & OBrien, E. (2015). Spatial time-dependent reliability analysis of reinforced concrete slab bridges subject to realistic traffic loading. *Structure and Infrastructure Engineering*, 1–15. doi: 10.1080/15732479.2015.1086385.

Honshu Shikoku Bridge Authority. (1980). Superstructure design standard. (In Japanese), Tokyo.

Ministry of Communications and Transportation. (2004). *General code for design of highway bridges and culverts.* JTG D60-2004, Beijing (In Chinese: China Communications Press).

Nowak, A., Lutomirska, M., & Ibrahim, F. (2010). The development of live load for long span bridges. *Bridge Structures: Assessment, Design and Construction, 6*, 73–79.

Nowak, A. S., & Rakoczy, P. (2013). WIM-based live load for bridges. *KSCE Journal of Civil Engineering, 17*, 568–574.

O'Brien, E. J., Hayrapetova, A., & Walsh, C. (2012). The use of micro-simulation for congested traffic load modeling of medium-and long-span bridges. *Structure and Infrastructure Engineering, 8*, 269–276.

O'Brien, E. J., Schmidt, F., Hajializadeh, D., Zhou, X. Y., Enright, B., Caprani, C. C., & Wilson, S. (2015). A review of probabilistic methods of assessment of load effects in bridges. *Structural Safety, 53*, 44–56.

O'Brien, E. J., Lipari, A., & Caprani, C. C. (2015). Micro-simulation of single-lane traffic to identify critical loading conditions for long-span bridges. *Engineering Structures, 94*, 137–148.

O'Brien, E. J., Schmidt, F., Hajializadeh, D., Zhou, X. Y., Enright, B., Caprani, C. C., & Wilson, S. (2015). A review of probabilistic methods of assessment of load effects in bridges. *Structural Safety., 53*, 44–56.

O'Connor, A., & Eichinger, E. M. (2007). Site-specific traffic load modelling for bridge assessment. *Proceedings of the ICE-Bridge Engineering, 160*, 185–194.

Pelphrey, J., Higgins, C., Sivakumar, B., Groff, R. L., Hartman, B. H., Charbonneau, J. P., … Johnson, B. V. (2008). State-specific LRFR live load factors using weigh-in-motion data. *Journal of Bridge Engineering, 13*, 339–350.

Ruan, X., & Zhou, K. P. (2014). *Vehicle characteristics and load effects of four-lane highway.* 7th International Conference of Bridge Maintenance, Safety and Management, Shanghai: Taylor and Francis.

Ruan, X., Zhou, X. Y., & Guo, J. (2012). Extreme value extrapolation for bridge vehicle load effect based on synthetic vehicle flow. *Journal of Tongji University: Natural Science., 40*, 1458–1485. (In Chinese).

Ruan, X., Zhou, J. Y., & Yin, Z. Y. (2014). *Concepts of developing traffic load model for multi-span cable supported bridges* (pp. 132–140). ASCE International Conference on Sustainable Development of Critical Infrastructure, Shanghai.

Virlogeux, M. (1999). Recent evolution of cable-stayed bridges. *Engineering Structures, 21*, 737–755.

Virlogeux, M. (2001). Bridges with multiple cable-stayed spans. *Structural Engineering International, 22*, 27.

Wu, J., Yang, F., Han, W., Wu, L., Xiao, Q., & Li, Y. (2015). Vehicle load effect of long-span bridges: Assessment with cellular automaton traffic model. *Transportation Research Record: Journal of the Transportation Research Board, 2481*, 132–139.

Computer vision-based displacement and vibration monitoring without using physical target on structures

Tung Khuc and F. Necati Catbas

ABSTRACT

Although vision-based methods for displacement and vibration monitoring have been used in civil engineering for more than a decade, most of these techniques require physical targets attached to the structures. This requirement makes computer vision-based monitoring for real-life structures cumbersome due to need to access certain critical locations. In this study, a non-target computer vision-based method for displacement and vibration measurement is proposed by exploring a new type of virtual markers instead of physical targets. The key points of measurement positions obtained using a robust computer vision technique named scale-invariant feature transform show a potential ability to take the place of classical targets. To calculate the converting ratio between pixel-based displacement and engineering unit (millimetre), a practical camera calibration method is developed to convert pixel-based displacements to engineering unit since a calibration standard (a target) is not available. Methods and approaches to handle challenges such as low contrast, changing illumination and outliers in matching key points are also presented. The proposed method is verified and demonstrated on the UCF four-span bridge model and on a real-life structure, with excellent results for both static and dynamic behaviour of the two structures. Finally, the method requires a simple, less complicated and more cost-effective hardware compared to conventional displacement and vibration monitoring measuring technologies.

1. Introduction

Demand from infrastructure systems has been expanding to meet societal needs for economic prosperity and life quality. For example, highway bridges need to accommodate more and heavier traffic with more axle weights to transport goods and people. Thus, restoring and improving the infrastructure systems to ensure the safety and reliability is very important, which has been pointed out as one of the Grand Challenges of Engineering in the twenty-first century (The National Academy of Engineering, 2008). Although assessment of the infrastructure systems (e.g. bridges, tunnels, roads and pipes) is conducted via periodical visual inspection traditionally, utilising structural health monitoring (SHM) techniques for managing infrastructure assets has recently received more attention from researchers and civil infrastructure owners (Catbas & Kijewski-Correa, 2013). With maturing sensing technologies, reasonable prices and the developments in computer-based data analyses, the SHM implementations have shown opportunities to complement the traditional visual inspection methods (Çatbaş, Kijewski-Correa, & Aktan, 2013; Zaurin & Catbas, 2010a).

Reliably obtaining structural responses and tracking them for decision-making purposes is the first critical step for SHM. A change in dynamic and/or static response trend of a structure would be an indicator of damage occurring on the structure or some other structural issues that need to be evaluated. Most fundamental and common responses employed in SHM are acceleration, strain, tilt, displacement since these can clearly reflect both local and global behaviours of an existing structure under various loading conditions. Moreover, out of these common response types, displacement is arguably the most important one as the most developed performance-based design is direct displacement-based design where performance is related to acceptable damage and damage to displacement. As such displacement can be directly used for safety and serviceability limit state estimation despite displacement poses a particular measurement challenge due to reference requirement. Motivated by those reasons, this paper is aimed to develop a completely contactless, cost-effective and practical displacement measuring method for real-life structures where displacement monitoring might not be easy or possible.

Traditionally, displacement sensors such as linear variable differential transformers (LVDTs), slide wire potentiometer or dial gauges have been utilised to collect displacement response. These classical sensors are quite convenient to use in laboratories; however, are not practical to deploy on a real-life structure due to several reasons such as the need for stationary platforms near measurement points to mount sensors, and limitation of the sensor range. There are other proposed approaches to tackle those drawbacks of classical sensors such as Global Position System

(GPS), Interferometric Radar, Laser Doppler Vibrometer and Scanning Laser Vibrometer. Although radar- and laser-based methods provide high precision, they require very high-cost equipment. Currently, the GPS system costs are coming down, however, the limitations due to GPS accuracy as well as possible sampling rates remain as issues to be solved.

Those limitations make the GPS system be commonly suitable for specific applications such as cable-bridge monitoring studies due to those structures have larger displacement range and low natural frequencies (Im, Hurlebaus, & Kang, 2011). To address all above limitations, vision-based monitoring has been exploded due to its practical deployment and cost-effectiveness (Catbas, Zaurin, Gul, & Gokce, 2012; Zaurin & Catbas, 2010b). Regarding vision-based displacement and vibration measurement, several studies proposed algorithms for determining deflection and vibration from multi-points on a small beam by means of matching detected edges or markers between consecutive image frames collected by a digital camera (Cantatore, Cigada, Sala, & Zappa, 2009; Jurjo, Magluta, Roitman, & Gonçalves, 2010; Patsias & Staszewskiy, 2002; Poudel, Fu, & Ye, 2005; Rucka & Wilde, 2005; Shi, Xu, Wang, & Li, 2010; Sładek et al., 2013). Even though most of these studies could obtain both static deflection and dynamic vibration of a beam, the algorithms were limited for laboratory implementations. This is due to the fact that the studies were only practical for a small structure; hence, the entire structure could be taken inside an image view.

Conducting SHM studies for real-life structures using vision-based techniques has been explored by some researchers due to practical nature of the measurements. In 2003, (Jauregui, White, Woodward, & Leitch, 2003) conducted a series of tests in New Mexico State to measure displacements of targets attached under main girders of several bridges by identifying their three-dimensional (3-D) locations. By obtaining images of targets at different viewpoints, the authors successfully determined 3-D locations of targets by utilising the principle of triangulation algorithm (a basic computer vision technique). Another research to determine displacements of a real structure was implemented in 2006 at a steel bridge in Korea (Lee & Shinozuka, 2006). In this study, the authors developed a practical vision system that could obtain displacement data at real-time speed. Using a special target containing four black dots on white background, these dots could be detected in terms of colour filtering for tracking their motions in time domain. Moreover, the pre-defined distances among the dots had been utilised for converting from pixel unit to engineering unit (millimetre).

That approach was later improved to obtain displacements from multiple locations of structures by synchronising numerous vision systems with a wireless network (Lee, Fukuda, Shinozuka, Yun, & Cho, 2007). Lately, application of normalised cross-correlation (NCC) imaging algorithm has become quite popular for vision-based displacement measurement methods. By calculating correlation of region of interest (ROIs) of two target images, movements of the ROIs between consecutive imaging frames were determined even at sub-pixel accuracy. Although this approach could not be executed at a real-time speed, very small displacements and precise dynamic characteristics of structures could be obtained. The method was successfully conducted on laboratory experiments and on several bridges in Korea and Hong Kong (Kim, Lee, Kim, & Kim, 2013; Ye et al., 2013). Some targetless practices have also been implemented by utilising the NCC imaging algorithm on natural textures of structure images. However, only pixel-based vibration for identifying structural characteristics was interested during these experiments (Kim, Jeon, Kim, & Park, 2013). Additionally, inconsistent results were observed due to the effects of low contrast on structure textures (Busca, Cigada, Mazzoleni, Tarabini, & Zappa, 2013).

2. Challenges, research goals and contributions

A general framework for conducting a vision-based displacement measurement includes: (i) Capturing video clips of targets attached on monitored locations using an extra-tele lens camera, (ii) Identifying dominant regions on targets by means of image processing, (iii) Determining image features of these dominant regions to match them between consecutive frames and (iv) Calculating pixel-based displacements; and then converting them to the engineering unit (millimetre) using standards on the targets (e.g. pre-defined shape dimensions). Following these steps, a target plays a very important role since it is a sort of dominant marker for identifying and tracking while processing images as well as a benchmark for converting pixel-based displacements to millimetre-based displacements. However, utilising target attachment on real-life structures would be a challenge similar to mounting of conventional sensors on the structure. This fact makes conventional vision-based displacement monitoring not a fully non-contact deployment, which may limit its widespread use.

The goal of this research is to further improve displacement monitoring by developing a non-target vision-based method that will address the limitation of target attachment. To discard the physical targets in general vision-based displacement monitoring framework, three objectives are proposed as follows:

(1) Exploration of a new type of virtual marker on measurement locations called imaging key points that can replace conventional physical targets
(2) Development of a conversion method based on the camera calibration technique to transfer pixel-based displacements to engineering-based (millimetre) displacements since physical targets no longer exist.
(3) Proposing methods and approaches to handle challenges such as low contrast, changing illumination and outliers in matching key points

By achieving fully non-contact monitoring, implementing the proposed method will be more practical, especially for real-life structures. Without using the target attachment in a vision-based displacement monitoring, most of the field works and requirements such as installing targets, sensors and DAQ systems, as well as wiring cables are not needed anymore. That improvement enables not only a cost-effective measurement method, but also a possibility to obtain structural responses from difficult access locations. With the implementation advantages plus the generic response that can be obtained, the method gives an opportunity for developing a more comprehensive and practical SHM framework.

The proposed framework is validated on a four-span bridge model at the University of Central Florida (UCF) Structures Laboratory, subsequently, a field verification is conducted on an elevated guideway structure. The theoretical background is presented in Section 3 while the laboratory and real-life verifications of the proposed method are discussed in Sections 4 and 5, respectively.

3. Theoretical background

3.1. Key points extraction as new virtual markers

Image matching is a fundamental aspect of many problems in computer vision including object or scene recognition, rebuilding 3-D structure, stereo and motion tracking (Lowe, 2004). To match different images of the same object, researchers commonly extract image features of the object that are invariant to such as image translation, rotation, scaling and illumination changing. The image matching technique utilises a general procedure for vision-based displacement monitoring while the matched and tracked objects are measurement positions. Traditionally, previous studies for vision-based displacement measurement employ physical targets as a type of pre-defined image feature; consequently, target attachment is a requirement. A target including known-dimension circles or rectangles provides dominant markers (e.g. centre and/or corner points of those geometrical shapes) for matching and tracking easily by means of fundamental image processing algorithms. In this paper, due to the shortcomings of using target attachment as mentioned in the previous sections, image key points (a natural image feature) are used as virtual markers of measurement locations that replacing any physical targets.

In computer vision field, a key point is defined as a special pixel that has dominant textures or characteristics comparing to its neighbours. For example, the key points can be obtained as corner points of a checkboard (Figure 1) by implementing the Harris corner detection (Harris & Stephens, 1988). Even though there are different types of key points obtained by different computer vision algorithms, the robust key points are interested due to their invariance, reliability and consistency. Once the robust key points on monitoring positions are detected, motions of these structural locations will be determined in terms of the key points movements, which can be tracked across consecutive image frames.

In the pool of extracting robust key points algorithms, following methods have been acknowledged as the most robust techniques including scale invariant feature transform (SIFT) (Lowe, 2004), speed-up robust feature (Bay, Ess, Tuytelaars, & Van Gool, 2008), binary robust invariant scalable key points (Leutenegger, Chli, & Siegwart, 2011) and fast retina key points (Alahi, Ortiz, & Vandergheynst, 2012). As one of the most comprehensive and popular usage algorithms, the SIFT method is utilised in this study for obtaining key points as well as matching them across measurement position image sequence.

Following the SIFT method, an input image $I(x, y)$ is filtered using the Gaussian kernel to discard noise that commonly dominate key point candidates. Since it is impossible to find the most suitable Gaussian kernel, a scale-space of Gaussian functions corresponding to different standard deviation values of σ_i,

Figure 1. Key points (red dots) as corner points of a checkboard.

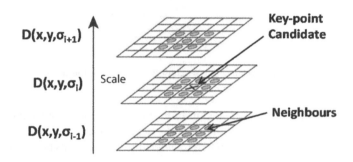

Figure 2. Key point identification using the local extrema detection algorithm – Modified from a figure in Lowe (2004).

namely $G(x, y, \sigma_i)$, is utilised to scan all potential candidates of key points at every scale of filtering. The scale-space of filtered images $L(x, y, \sigma_i)$ is derived by:

$$L(x, y, \sigma_i) = G(x, y, \sigma_i) \otimes I(x, y) \quad (1)$$

where $\otimes$ called the convolution, a mathematical operation applied for two matrices, and:

$$G(x, y, \sigma_i) = \frac{1}{2\pi\sigma_i^2} e^{-(x^2+y^2)/2\sigma_i^2} \quad (2)$$

To efficiently detect key point locations, differences of two adjacent filtered images (e.g. filtered images $L(x, y, \sigma_i)$ and $L(x, y, \sigma_{i+1})$ have been determined in which the local extrema detection algorithm is then applied. Those differences of filtered images $D(x, y, \sigma_i)$ can be calculated as follows:

$$D(x, y, \sigma_i) = L(x, y, \sigma_{x+1}) - L(x, y, \sigma_i) \quad (3)$$

Subsequently, a key point can be detected at the location which has a local extrema value on the differences of filtered images $D(x, y, \sigma_i)$. That local extrema detection process is to compare the candidate intensity value (marked with X) to its 26 neighbours in 3×3 regions at the current and adjacent scales (marked with circles) as shown in Figure 2. Consequently, the key points found from the previous step have been tested their robustness to reject the low contract and the poor location (e.g. along an edge) candidates. The detailed explanation as well as all related equations can be found in (Lowe, 2004).

Figure 3 illustrates the detected key points from two consecutive image frames of a monitored region. It is seen that many key

Figure 3. Detected key points from two consecutive images by the SIFT algorithm.

points have been identified in both images (843 key points and 804 key points, respectively). In the next section, these two sets of key points will be matched to determine movements between them in pixel-based unit.

3.2. Matching key points for determining displacements of measurement location

In general, obtained key point sets from two consecutive images are matched in terms of comparing their descriptor vectors. Following the SIFT method, a descriptor vector of a key point is built up from the gradient magnitudes and orientations of the key point neighbours, which can be calculated as follows:

$$M(x, y) = \sqrt{\left(L(x+1, y) - L(x-1, y)\right)^2 + \left(L(x, y+1) - L(x, y-1)\right)^2}$$
(4)

$$\theta(x, y) = \tan^{-1}\left(\frac{L(x, y+1) - L(x, y-1)}{L(x+1, y) - L(x-1, y)}\right)$$
(5)

where $m(x, y)$ is the gradient magnitude, and $\theta(x, y)$ is the orientation of each neighbouring pixel. A key point on a certain image (e.g. image kth on a video sequence) will be matched to another key point on the next image $(k + 1)$th by identifying its nearest neighbour. The nearest neighbour can be obtained by determining the minimum Euclidean distance between the key point descriptor vector on the image kth and all other key point descriptor vectors from the image $(k + 1)$th.

All images obtained from a video clip capturing a measured location are processed to extract key points. Displacements of the measured location are then determined by matching the key points between the consecutive images and calculating pixel-based distance between them. Let S_k and S_{k+1} are the two sets of the matched key points corresponding to two consecutive images kth and $(k + 1)$th, so that:

$$S_k = P_k^1, P_k^2, \dots P_k^{n-1}, P_k^n; \text{ and}$$
$$S_{k+1} = P_{k+1}^1, P_{k+1}^2, \dots P_{k+1}^{n-1}, P_{k+1}^n$$
(6)

where n is the number of key points in each matched set; and P is the key points. The centroid point in each set S_k and S_{k+1} can be found based on locations of the key points called $\left(X_k^c \ Y_k^c\right)$ and $\left(X_{k+1}^c \ Y_{k+1}^c\right)$, respectively, as illustrated in Figure 4. The distance between two sets S_k and S_{k+1} is calculated in pixel unit:

$$d_k^X = (X_k^c - X_{k+1}^c); \text{ and}$$
$$d_k^Y = (Y_k^c - Y_{k+1}^c)$$
(7)

An observed problem when calculating the distance d_k is false-matches from the SIFT outcomes. Even though the algorithm is acknowledged as one of the most robust methods, the false-matches occurred at a rate of 10% in a good, laboratory light condition; and up to a ratio of 30% as images were captured under bad light condition. In the case worst scenario, even more than 60% of the matches may not be correct if textureless objects were acquired under changing light illumination. To address this issue, the trimmed mean algorithm as a statistical measure technique, is utilised to discard the percentage outliers by rejecting the matches that have the lowest or the highest distances determined from the SIFT outcomes. Since the percentage in the trimmed mean algorithm is a parameter, it must be adjusted corresponding to a particular monitoring condition.

For example, the percentage value of outlier will be small for the case of good image quality and vice versa. In this study, this parameter is manually determined by gradually increasing its value (e.g. 10, 15, 20% and so on) until the displacement results corresponding to consecutive parameter values are stabilised. The result of matches between two consecutive images after discarding 30% of outliers (the lowest 15% and the highest 15% of the distances) is illustrated in Figure 4. The procedure will be repeated for all images along the video clip in time domain;

Figure 4. Matching key points of two consecutive images after discarding 30% outliers.

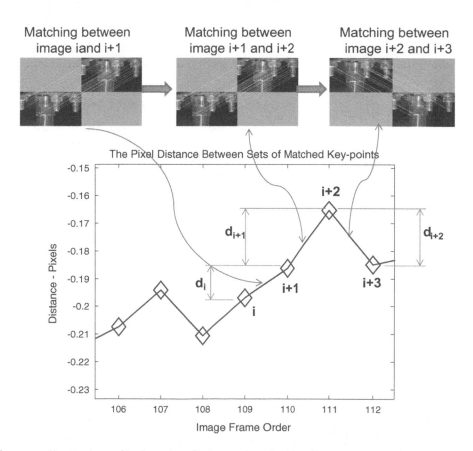

Figure 5. Displacements of a measured location by matching key points of its images along the video clip.

and displacements of the measured locations in time history is presented in Figure 5.

3.3. Camera calibration for determining engineering displacements from pixel-based measurement

Apart from using a physical target as a marker, an attached target can also be used as a conversion standard to transfer obtained displacements from pixel-based unit to millimetre-based unit in

conventional vision-based displacement measurements. In this study, since a physical target no longer exists, an alternative approach is employed to determine conversion by utilising camera calibration to obtain absolute deflection measurement. Theoretically, the relationship between the world distance D and the image distance h can be explained in Figure 6 and be derived via:

$$\frac{h}{D} = \frac{f}{Z} \tag{8}$$

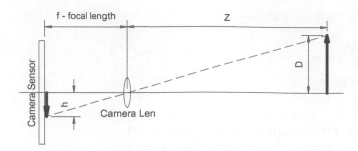

Figure 6. Camera theory.

Figure 8. The four-span bridge at the UCF structural laboratory.

where f is the focal length of a camera; Z is the distance from the camera to the object D (measurement location). In the image coordinate, the distance h is measured in pixel-based unit as shown below:

$$h = d \times p \quad \text{(pixel)} \tag{9}$$

where d is the object image in the pixel-based unit (pixel); p is the unit length of camera sensor (mm/pixel), which may be provided by the camera manufacturer.

Substituting Equation (9) into Equation (8) and rearranging gives Equation (10), in which the converting ratio R is demonstrated as an inversely proportional function to Z value:

$$R = \frac{d}{D} = \frac{f}{p \times Z}\left(\frac{\text{pixel}}{\text{mm}}\right) \tag{10}$$

Generally, the ratio $\frac{f}{p}$, an internal parameter of a camera, can be calculated via camera specifications provided by its manufacturer; however, this information can hardly be found for the majority of low-cost cameras. Due to this reason, the ratio $\frac{f}{p}$ is determined using a camera calibration algorithm. In this study, a Samsung HMX-F90 camcorder is calibrated using a checkerboard including 8×8 squares with dimension of 56.4×56.4 mm as illustrated in Figure 7. First, the checkerboard is captured by the camera; and then, some reference points are detected in both the image and world coordinates to determine the distances among the reference points in both coordinates d_{cal} and D_{cal}. Consequently, the converting R_{cal} can be calculated by Equation (10). At a capturing distance of 1.83 m, the converting ratio R_{cal} of the camera is 21.91 pixels/mm. The ratio $\frac{f}{p}$ is then calculated

for this particular camcorder (Samsung HMX-F90); and the converting ratio R function is written as follows:

$$R = \frac{40.095}{Z}\left(\frac{\text{pixel}}{\text{mm}}\right) \tag{11}$$

Other cameras can also be calibrated following this procedure; and it is a one-time implementation for every camera.

4. Laboratory verification

The proposed computer vision-based monitoring demonstrated first on a small-scale bridge at the UCF Structures Lab. The bridge named UCF four-Span Bridge Model consists of two 304.8 cm main continuous spans and two 120 cm approach spans. The model deck is 120 cm wide 3.18 mm steel sheet compositely connected to two HSS 25 mm $\times$ 25 mm $\times$ 3 mm steel girders separated 60.96 cm from each other as shown in Figure 8. As a part of various SHM studies at the UCF, there are a large number of sensors attached on the model such as strain gages, accelerometers, LVDTs, tiltmeters and fibre Bragg grating sensors.

Moreover, some common bridge damage scenarios can be replicated by changing boundary conditions at the supports or altering the local stiffness of the girders by losing some bolts to reduce composite action of the model elements. To simulate traffic loading, small-scale vehicles are deployed back and forth on

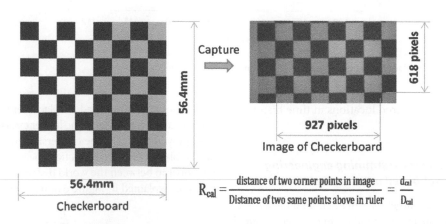

Figure 7. Camera calibration for converting pixel unit to millimetre unit.

Table 1. The test specifications.

	Speed[a] (m/s)	Weight[a] (kg)	Distance[b] (m)	Angle[b] (°)
Test 1	.32	12.3	2.04	0
Test 2	.26	12.3	3.74	0
Test 3	.27	12.3	3.75	6
Test 4	.30	12.3	2.07	11
Test 5	.79	12.3	2.04	0
Test 6	.37	5.30	2.04	0

[a]Speed and weight of the small-scale vehicle.
[b]Distance value Z from camera to the measured location; and angular orientation of camera.

the bridge deck. In this section, displacements of the small-scale bridge due to traffic-induced loading from a small-scale vehicle are obtained by utilising the non-target vision-based displacement measurement method. The raw displacement data of the experiments are verified using pre-attached LVDTs. Since the camera can capture images with the rate of 30 frames per seconds, the displacements of the structure are also acquired with the same sample rate at 30 Hz. Thus, the natural frequencies of the structure are identified from the dynamic component of the displacement data and then compared to the structural frequencies identified from accelerometer data.

4.1. Experiment design

For verification purposes, the monitored region is selected as close as possible to the location of pre-attached sensors (LVDT and accelerometer) located under the main girder at the two-fifth span point. There is a data acquisition system for simultaneously collecting data from all sensors and the camera. To confirm reliability and consistency of the proposed method, a total of six (6) tests are conducted by altering small-scale vehicle weights and speeds, as well as locations and angular orientations of the camera as detailed in Table 1. A small-scale vehicle is driven over the bridge deck following pre-defined loading configurations and speeds. As the small-scale vehicle is actually a dynamic load, it induces not only static displacements, but also dynamic vibrations on the structure. Such obtained displacements and vibrations are presented and verified in the following sections (Figure 9).

4.2. Laboratory results and discussion

The displacement results at the measurement location of all tests are shown in Figure 10. The sample rates of all data-sets measured by both LVDT and the proposed vision based method are at 30 Hz. In each of the graphs, both raw displacement data collected from the LVDT and the proposed vision-based method are synchronised and illustrated in a comparative fashion.

It is seen that although the experimental set-ups are different among the tests, displacement values determined from the non-target vision-based method highly correlate with the results obtained from the LVDT sensor. This observation is even confirmed strongly at every peak of the vibration data as shown in the insets in Figure 10. To measure the correlation behaviour between two data-sets, the correlation coefficient factors (ρ) are determined by Equation (12) for every experiment. In addition, a statistical measure named determination coefficient (R^2) factors is computed to determine how well the two data-sets match together. The R^2 value can be calculated using Equation (13):

$$\rho = \frac{\left| \sum_i \left(d_L(i) - \mu_{d_L} \right) \times \left(d_v(i) - \mu_{d_v} \right) \right|}{\sqrt{\sum_i \left(d_L(i) - \mu_{d_L} \right)^2} \sqrt{\sum_i \left(d_v(i) - \mu_{d_v} \right)^2}} \quad (12)$$

where d_L and d_v are the dynamic displacement values extracted by filtering out the static component of the raw data from the LVDT sensor and the proposed vision-based method, respectively; and μ_{d_L} and μ_{d_v} are the mean values of two above data-sets. The values of ρ vary from .0 to 1.0; and $\rho = 1.0$ shows perfect correlation, whereas $\rho = .0$ indicates no correlation between two data-sets. The dynamic data are used to measure correlation at every peak. If raw data are used, the correlation factor will be 1.0 or .99999:

$$R^2 = 1 - \frac{\sum_i \left(r_v(i) - r_L(i) \right)^2}{\sum_i \left(r_v(i) - \mu_{r_v} \right)^2} \quad (13)$$

where r_L and r_v are the raw displacement values obtained by the LVDT and proposed vision-based method, respectively; and μ_{r_v} is the mean value of the raw data-set determined by the proposed

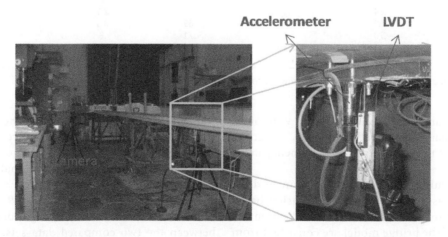

Accelerometer   LVDT

Figure 9. Measured location and experimental set-up.

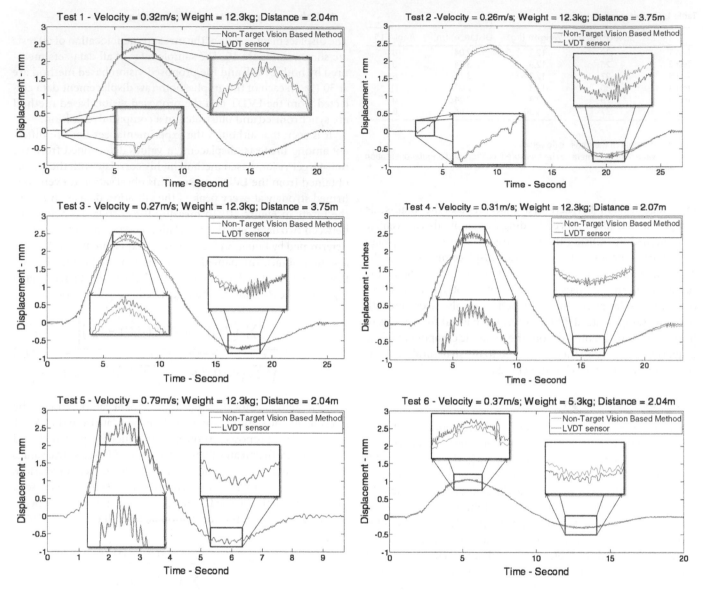

Figure 10. Comparison of displacement results by using the LVDT and the proposed method.

Table 2. Correlation and error analysis for the proposed method.

	Vision-based method		LVDT		Error			
	Max Disp.[a] (mm)	Min Disp.[a] (mm)	Max Disp.[a] (mm)	Min Disp.[a] (mm)	Max Disp.[a] (%)	Min Disp.[a] (%)	Corr. (ρ)	R^2
Test 1	2.45	−.71	2.47	−.72	.8	1.4	.969	.9997
Test 2	2.36	−.64	2.41	−.67	2.1	4.5	.971	.9998
Test 3	2.37	−.68	2.45	−.69	3.3	1.4	.961	.9995
Test 4	2.46	−.70	2.48	−.71	.8	1.4	.929	.9995
Test 5	2.51	−.74	2.50	−.74	.4	.0	.951	.9987
Test 6	1.06	−.30	1.07	−.31	.9	3.2	.938	.9988

[a]The static maximum and minimum displacement values after filtering dynamic behaviour.

method. The values of R are from .0 to 1.0; and the R value of 1.0 implies the perfect similarity between two data-sets. The entire data are used to measure the similarity of displacement pattern (shape).

The comparison results between two displacement data-sets obtained from a classical sensor and a new vision-based method are described in Table 2, maximum and minimum static displacements of the bridge model are consistent from Test 1–5 based on the responses from the same small-scale

vehicle weight (12.3 kg). However, the lighter weight vehicle (5.3 kg) being utilised at Test 6 induces smaller displacement amplitudes and the ratio of displacement amplitudes is similar to the weight ratio. It is also seen that the correlation coefficient ρ values (from .929 to .971) and the determination coefficient R-squared values (from .9987 to .9998) are close to 1.000, which indicate a very high correlation and similarity between the two compared data-sets. Hence, the displacements obtained by the non-target displacement measurement

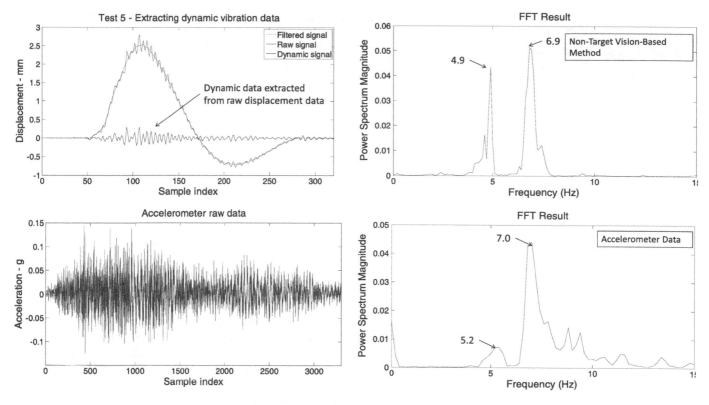

Figure 11. Comparison of identified natural frequencies using the proposed method and the accelerometer data.

Figure 12. (a) Trains running on the viaduct, and (b) runway under the viaduct.

method show comparable and accurate results when under laboratory condition.

Furthermore, the dynamic characteristics of the four-span bridge model can be captured by analysing the dynamic component of the raw displacement data-sets. By FFT-transforming the dynamic displacements from time domain to frequency domain, the natural frequencies of the bridge model can be identified as shown in Figure 11 at 4.9 and 6.9 Hz. In the meantime, vibration response of the bridge model is collected using an accelerometer attached at the same location captured by the camera. Figure 11 also depicts the raw data recorded using that accelerometer; from which the first and second natural frequencies of the structure can be detected. The identified frequencies from accelerometer data are 5.2 and 7.0 Hz that closely match to the frequencies of 4.9 and 6.9 Hz identified using the proposed vision-based method. However, in spite of the good match for the first two frequencies,

higher frequencies of the bridge structures can only be identified using accelerometer data.

Through the results of the non-target vision-based displacement measurement method described in this section, it is seen that the proposed algorithm cannot only obtain static displacements, but also identify dynamic characteristics of the four-span bridge model. The outcomes from different experimental set-ups confirm the accuracy of the proposed method consistently in the laboratory. For further verification, a real-life implementation is conducted and the corresponding results are presented and discussed in the following section.

5. Field verification on an elevated guideway

An automated people mover (APM) system mainly includes four basic components: trains, guideways, stations and a control

Figure 13. Measurement location and camera set-up.

system. For the system presented here, trains are designed to run on a viaduct system comprised of multiple span bridges that connect four airsides to the landside terminals as shown in Figure 12. Operating every 2 min from 5am to mid-night, maintenance of the skybus system is critical to guarantee the safety of passengers.

As a part of the APM guideway rehabilitation project, some spans of the viaduct are monitored to verify the behaviour of the structure after retrofitting. Because of the access and space limitation (Figure 12(b)), all monitoring equipment are designed wirelessly including accelerometers, strain gages and cameras in order not to interfere with the busy under viaduct traffic. Although there are different types of sensor data available from this monitoring project, only displacement and vibration responses of the viaduct are obtained using the proposed vision-based method under train loadings and the results are verified using sensor-based measurements.

5.1. Considerations for changing ambient illumination

In general, the most common obstacles that affect both the quality of images and the accuracy of obtained results when conducting a vision-based monitoring are (1) far distance from the camera to the measurement locations and (2) changing ambient illumination when collecting video clips. In this monitoring study, especially changing illumination issues had to be resolved. For the set-up of the experiment, the camera is focused on a measurement location near the mid-point of a main girder from a distance of 11.5 m away from the camera location, where a wireless accelerometer (AS1-ACC25) is attached (Figure 13).

Image sequence of the measurement location is captured using the camera when trains pass over the monitored span. Highly changing of ambient illumination and low contrast of images due to shade of the trains on the measurement location are the challenges to be solved (Figure 14). It is difficult to address these problems using common vision-based displacement measurement methods, which are based on the image correlation algorithms. Besides, the correlation approach is very sensitive to both illumination changes and low contrast of photography. By utilising the proposed method, the matching key points algorithm automatically selects the strongest key points from an appropriate illumination region to match. In this monitoring study, the measurement locations are captured under highly changing of the light illumination as well as dark condition (due to being under the bridge deck), the false-matching rate is a high value. Then, the percentage parameter of the trimmed mean method (the outlier discarding algorithm) is used and selected the value of 65% for successfully obtaining the best matches, as illustrated in Figure 15.

5.2. Obtaining Structural Displacements and Identifying Dynamic Frequencies

Figure 16 shows the displacements of the measurement location beginning from the time of the trains approaching the monitored

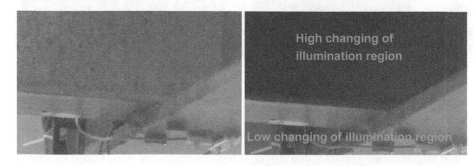

Figure 14. Highly changing illumination among image frames due to shade of the trains on the measurement positions.

(a)   (b)

Figure 15. (a) Automatically matching the strongest key points for low change of illumination region, and (b) the best matches after utilising the outlier discarding algorithm.

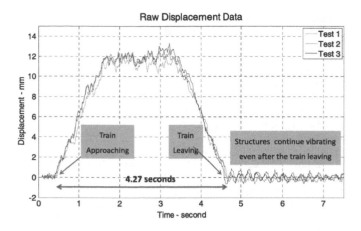

Figure 16. Raw displacement data of a measurement location on obtained using the proposed method.

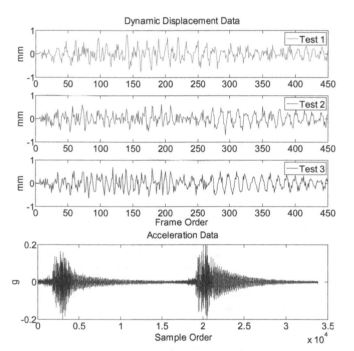

Figure 17. Dynamic responses of the main girder extracted by the proposed method and an accelerometer (AS1-ACC25).

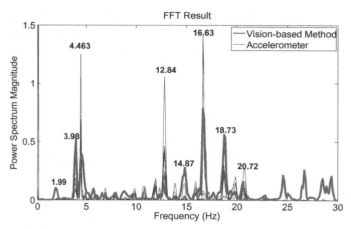

Figure 18. Comparison of natural frequencies of the girder identified by the proposed vision-based method and accelerometers.

Table 3. Comparison of natural frequencies.

	Vision-based (Hz)	Accelerometer (Hz)	Error (%)
1st Freq	3.98	3.97	.18
2nd Freq	4.57	4.46	2.44
3rd Freq	12.76	12.84	.62
4th Freq	14.87	14.79	.54
5th Freq	16.62	16.63	.06
6th Freq	18.73	18.77	.21
7th Freq	20.61	20.72	.53

span to the time of their exit. A total of three data tests were collected that illustrate the consistency of the general behaviour such as the deflection pattern, maximum displacements (~12 mm) and the response time (~4.27 s). Although there is not any LVDT information to verify the maximum displacements in a comparative sense, the deflection pattern and the time of structural response are observed to be similar to strain pattern and response time (~4.32 s) acquired from a strain gage mounted at the same location.

Modal frequencies of the main girder can be identified using the dynamic component of the displacement data extracted from the raw displacement history by filtering out the static response. In addition, acceleration responses of the same girder are collected by total of 10 accelerometers attached at different locations on the girder (including the one at the same camera capturing location, AS1-ACC25). Dynamic responses from

both displacement and acceleration data are shown in Figure 17. Natural frequencies found via the processing of both camera and accelerometer data are presented in Figure 18. While one can observe some minor differences in the identified frequencies, a considerable number of the dominant frequencies identified using two different approaches match. The errors between two sets of frequencies are calculated in Table 3 indicating errors less than 2.5%. The comparison result confirms that the proposed vision-based method works efficiently even in case of a real-life structure under difficult lighting condition.

6. Conclusions

A new framework for vision-based displacement measurement has been proposed in this paper. The new method is able to measure both static and dynamic displacements time histories as other state-of-the-art vision-based methods, but target attachment is not required. The study goal has been achieved by proposing a new type of virtual markers instead of using physical targets attached on the structure. The key points of measured locations are detected by means of a robust computer vision technique named SIFT. Low contrast, changing illumination and outliers in matching key points are critical challenges. Methods and approaches to address these challenges are presented. Moreover, a practical implementation for camera calibration is explored to convert pixel-based displacements to engineering unit since a calibration standard (on a target) is not available.

The proposed approach is verified and demonstrated on the four-span bridge in the UCF Structural Laboratory and on a real-life structure. The displacements obtained by the proposed vision-based techniques illustrate close results to the LVDT sensors for all tests. Furthermore, dynamic data can be obtained from raw displacement data as the camera can capture images at a speed rate of 30 Hz (laboratory tests) and 60 Hz (real-life implementation). By FFT-transforming the dynamic data to frequency domain, natural frequencies of both the four-span bridge model and the guideway structures can be identified that match to the ones identified by using acceleration data-sets. The maximum difference between the proposed computer vision-based method and the conventional sensors is less than 5% for all tests presented in this study.

Since there is no need to attach targets, implementation of the method is very practical as being a fully contactless deployment. This advantage provides convenience, especially for implementing at real-life structures. Finally, the hardware of the overall camera system is proposed at the simplest level and is more cost-effective than classical monitoring systems using DAQs and sensors. To obtain good displacement results from this proposed method, any external effect that induces movements to the camera stand/tripod must be minimised. The solution is to use a sturdy tripod conjunction with an appropriate and stationary place for the camera system (including camera and tripod) on a solid ground. The proposed method provides very good results for the laboratory tests and promising results from the initial fieldwork. More detailed field studies where more measurements can be taken and compared will be carried out and will be presented in the future as more field results become available.

Disclosure statement

No potential conflict of interest was reported by the authors.

Funding

This work was supported by NSF Division of Civil, Mechanical and Manufacturing Innovation [grant number 1463493]; partially by Türkiye Bilimsel ve Teknolojik Arastirma Kurumu (TUBITAK) [grant number 2221] and the West Nippon Expressway Company Limited.

References

Alahi, A., Ortiz, R., & Vandergheynst, P. (2012). *Freak: Fast retina keypoint.* Paper presented at the computer vision and pattern recognition (CVPR), 2012 IEEE conference, Providence, RI.

Bay, H., Ess, A., Tuytelaars, T., & Van Gool, L. (2008). Speeded-up robust features (SURF). *Computer Vision and Image Understanding, 110,* 346–359. doi:http://dx.doi.org/10.1016/j.cviu.2007.09.014

Busca, G., Cigada, A., Mazzoleni, P., Tarabini, M., & Zappa, E. (2013). *Static and dynamic monitoring of bridges by means of vision-based measuring system.* Paper presented at the IMAC XXXII A conference and exposition on structural dynamics, Orlando, FL.

Cantatore, A., Cigada, A., Sala, R., & Zappa, E. (2009). Hyperbolic tangent algorithm for periodic effect cancellation in sub-pixel resolution edge displacement measurement. *Measurement, 42,* 1226–1232. doi:http://dx.doi.org/10.1016/j.measurement.2009.06.001

Catbas, F. N., & Kijewski-Correa, T. (2013). Structural identification of constructed systems: Collective effort toward an integrated approach that reduces barriers to adoption. *Journal of Structural Engineering, 139,* 1648–1652.

Çatbaş, F. N., Kijewski-Correa, T., & Aktan, A. E. (2013). *Structural identification of constructed systems.* Reston (VI): American Society of Civil Engineers.

Catbas, F. N., Zaurin, R., Gul, M., & Gokce, H. B. (2012). Sensor networks, computer imaging, and unit influence lines for structural health monitoring: Case study for bridge load rating. *Journal of Bridge Engineering, 17,* 662–670. doi:http://dx.doi.org/10.1061/(asce)be.1943-5592.0000288

Harris, C., & Stephens, M. (1988). *A combined corner and edge detector.* Paper presented at the Alvey vision conference (pp. 1–23), September 6, 1988. University of Manchester, Manchester, UK.

Im, S. B., Hurlebaus, S., & Kang, Y. J. (2011). Summary review of GPS technology for structural health monitoring. *Journal of Structural Engineering, 139,* 1653–1664.

Jauregui, D., White, K., Woodward, C., & Leitch, K. (2003). Noncontact photogrammetric measurement of vertical bridge deflection. *Journal of Bridge Engineering, 8,* 212–222. doi:http://dx.doi.org/10.1061//asce/1084-0702/2003/8:4/212

Jurjo, D. L. B. R., Magluta, C., Roitman, N., & Gonçalves, P. B. (2010). Experimental methodology for the dynamic analysis of slender structures based on digital image processing techniques. *Mechanical Systems and Signal Processing, 24,* 1369–1382. doi:http://dx.doi.org/10.1016/j.ymssp.2009.12.006

Kim, S. W., Jeon, B. G., Kim, N. S., & Park, J. C. (2013). Vision-based monitoring system for evaluating cable tensile forces on a cable-stayed bridge. *Structural Health Monitoring, 12,* 440–456. doi:http://dx.doi.org/10.1177/1475921713500513

Kim, S.-W., Lee, S.-S., Kim, N.-S., & Kim, D.-J. (2013). Numerical model validation for a prestressed concrete girder bridge by using image signals. *KSCE Journal of Civil Engineering, 17,* 509–517. doi:http://dx.doi.org/10.1007/s12205-013-0560-1

Lee, J. J., & Shinozuka, M. (2006). A vision-based system for remote sensing of bridge displacement. *NDT & E International, 39,* 425–431. doi:http://dx.doi.org/10.1016/j.ndteint.2005.12.003

Lee, J. J., Fukuda, Y., Shinozuka, M., Yun, C.-B., & Cho, S. (2007). Development and application of a vision-based displacement measurement system for structural health monitoring of civil structures. *Smart Structures and Systems, 3,* 373–384.

Leutenegger, S., Chli, M., & Siegwart, R. Y. (2011). *BRISK: Binary robust invariant scalable keypoints.* Paper presented at the computer vision (ICCV), 2011 IEEE international conference, Barcelona, Spain.

Lowe, D. G. (2004). Distinctive image features from scale-invariant keypoints. *International Journal of Computer Vision, 60,* 91–110.

National Academy of Engineering (2008). *Grand Challenges for Engineering.* National Academies Press, p. 56. Retrieved from www.engineeringchallenges.org

Patsias, S., & Staszewskiy, W. J. (2002). Damage detection using optical measurements and wavelets. *Structural Health Monitoring, 1,* 5–22. doi:http://dx.doi.org/10.1177/147592170200100102

Poudel, U. P., Fu, G., & Ye, J. (2005). Structural damage detection using digital video imaging technique and wavelet transformation. *Journal of Sound and Vibration, 286,* 869–895. doi:http://dx.doi.org/10.1016/j.jsv.2004.10.043

Rucka, M., & Wilde, K. (2005). Crack identification using wavelets on experimental static deflection profiles. *Engineering Structures, 28,* 279–288. doi:http://dx.doi.org/10.1016/j.engstruct.2005.07.009

Shi, J., Xu, X., Wang, J., & Li, G. (2010). Beam damage detection using computer vision technology. *Nondestructive Testing and Evaluation, 25,* 189–204. doi:http://dx.doi.org/10.1080/10589750903242525

Sładek, J., Ostrowska, K., Kohut, P., Holak, K., Gąska, A., & Uhl, T. (2013). Development of a vision based deflection measurement system and its accuracy assessment. *Measurement, 46,* 1237–1249. doi:http://dx.doi.org/10.1016/j.measurement.2012.10.021

Ye, X. W., Ni, Y. Q., Wai, T. T., Wong, K. Y., Zhang, X. M., & Xu, F. (2013). A vision-based system for dynamic displacement measurement of long-span bridges: algorithm and verification. *Smart Structures and Systems,* 363–379. doi:http://dx.doi.org/10.12989/sss.2013.12.3_4.363

Zaurin, R., & Catbas, F.N. (2010a). Integration of computer imaging and sensor data for structural health monitoring of bridges. *Smart Materials and Structures, 19,* 015019. doi:http://dx.doi.org/10.1088/0964-1726/19/1/015019

Zaurin, R., & Catbas, F. (2010b). Structural health monitoring using video stream, influence lines, and statistical analysis. *Structural Health Monitoring, 10,* 309–332. doi:http://dx.doi.org/10.1177/1475921710373290

Edgar Cardoso: a tribute to a brilliant bridge engineer

Paulo J. S. Cruz 🆔

ABSTRACT

Edgar Cardoso (1913–2000) was a brilliant bridge designer and an enthusiastic pioneer of experimental analysis of bridges and of the development and use of high-precision transducers and acquisition data systems to monitor the critical parameters affecting the structural behaviour. In his long and proficuous career, Professor Edgar Cardoso was very committed to the use of high-performance materials and to develop new structural concepts and innovative construction techniques. He was the designer of some outstanding audacious elegant bridges in several continents. Some of them were world records and deserved a worldwide diffusion and recognition. The holistic vision of such inspiring bridge engineer will be emphasised and part of its valuable legacy will be shared through a detailed presentation of several examples.

1. Introduction

Edgar António Cardoso de Mesquita was born in Porto, Portugal, on 11 May 1913. His father was military and Civil Engineer. At that time, Porto had two magnificent bridges that most probably inspired him to select civil engineering. The Maria Pia Bridge, designed by Théophile Seyrig & Gustave Eiffel and built by the Eiffel company (Figure 1) and the Luiz I Bridge, designed by Théophile Seyrig and built by Willebroeck (Figure 2). A detailed description of those two bridges can be found in Cruz and Lopes Cordeiro (2003).

He graduated in civil engineering at the Faculty of Engineering of the University of Porto in 1937 (Figure 3). Simultaneously he studied some years of electrical engineering that he recognised to be his passion. After the graduation, he integrated the Division of Bridges of the Junta Autónoma de Estradas, the Portuguese Roadway Administration, in Lisbon. In 1944, he became second-class engineer and, in 1951, first-class engineer. Simultaneously to his career in the public sector, he opened his office – Laboratory for Testing and Study of Structures and Foundations Eng. Edgar Cardoso – in 1944 (Instituto Superior Técnico [IST], 2001).

He also developed a brilliant academic and scientific career. On 22 December 1951, he took the place of professor of the bridges discipline at the Technical University of Lisbon for two years, becoming permanently appointed in 1953 (Figure 4). On 18 November 1960, he was admitted as a correspondent member of the Academy of Sciences of Lisbon. On 12 December 1968, he became an effective member of this academy (IST, 2001).

Edgar Cardoso's intuition has been progressively enhanced and purified with the experience he accumulated with hundreds of small-scale models and with the structural behaviour observation (Figure 5). He was ahead of his time and led the way in Portugal to the experimental analysis of structures and to the development and use of high-precision instruments for measuring the mechanical behaviour of reduced-scale models and bridges.

Edgar Cardoso designed some devices to automatically record influence lines of any effect of a resistant structure, resulting from the action of a force or system of forces that describes trajectories on its free surface (Figure 6). They have the capacity to trace on paper strip or on screen influence lines of moving forces (Figure 7).

Edgar Cardoso developed countless contact extensometers for all kinds of measurements and uses (IST, 2001). In the making process of his small-scale models, he used various types of materials (such as plaster, ebonite, cork agglomerate or acrylic), all of which had a common denominator: they are isotropic.

It can be stated that all significant works designed by Professor Edgar Cardoso had their dimensions tested in small-scale models. The experimental analysis of structures allows to Edgar Cardoso to go beyond the limitations of the 'analytic way'. When planning the Arrábida Bridge, Edgar Cardoso made a curious statement, very enlightening in terms of his preference for experimental methods (Cruz, 2014):

> I would frankly like to state that, although I always seek an analytical resolution for everything (sometimes at the cost of laborious calculations, as we have seen above), it was only once experimental confirmation was obtained that I proceeded to the algebraic calculation of the various issues. Only after the experimental analysis based on the tests on the reduced scale models did I formulate the various hypotheses that led to an analytical resolution of the

Figure 1. Maria Pia Bridge.

Figure 2. Luiz I Bridge.

Figure 3. Edgar Cardoso in the 1930s (Source: CRC Press).

Figure 4. Professor Edgar Cardoso at Technical University of Lisbon (in the 1950s) (Source: CRC Press).

various problems. We sometimes work backwards only to comply with what is classically established, why not confess it?

Another revolutionary invention by Edgar Cardoso in the field of photography was a panoramic camera (Figure 8). Based on the principle of making the camera rotate around the optical centre, by making the film move in the opposite direction to that of the rotation of the camera, so that the film which is being impressed may have instantaneous zero rotation speed in relation to the support point of the camera tripod. The result is the production of continuous panoramic negatives or slides without any vertical distortions, with 360°, 720° or as much as the length will allow. Thus, with a normal 36-print film, the same subject may appear up to four times on the same photographic film (Soares, 2003). For making the camera rotating he used a Volkswagen Beetle windscreen wiper motor.

2. Elegant and audacious bridges

2.1. Bridge over the Estuary of the Sousa River

This slender concrete arch bridge over the Estuary of Sousa River, near Porto, was built in 1948. This 115-m-span and 14.75-m-rise bridge (ratio 1/8) held the national record for the largest reinforced concrete arch (Figure 9). Including the viaducts, it has a total length of 165 m (Soares, 2003).

The arch is made up of two ribs, each one with a solid I-shaped section. They are linked by a brace. The deck is a light slab, the load of which is supported by the ribs by means of isolated piers. The falsework used was made in wood, with foundations in the supporting pinewood piles driven into the riverbed (Figure 10). In a certain way, this bridge can be considered an essay for the Arrábida Bridge over Douro River.

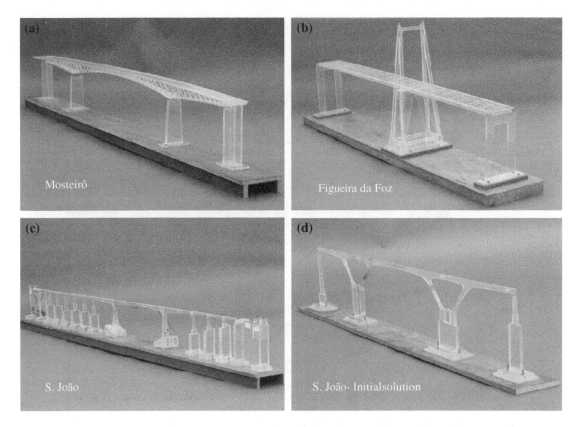

Figure 5. Small-scale models. (a) Mosteiró; (b) Figueira da Foz; (c) S. João; (d) S. João – Initial solution.

Figure 6. Mechanical electronic auto-influenciograph.

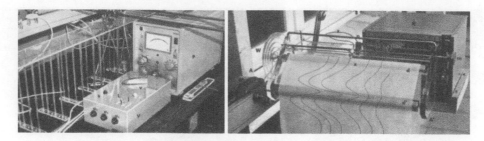

Figure 7. Auto-influenciograph tracing transversal influence lines and galvanometric recorder.

Figure 8. Panoramic camera and film projector in rotary mode (Source: CRC Press).

Figure 9. Bridge over Sousa River.

2.2. Abragão Bridge over the Tâmega River

It was inaugurated in 1949 and submerged in the waters of the Tâmega in 1988, when the dam of Torrão was concluded, state in which it is currently (Figure 11).

2.3. Santa Clara Bridge over the Mondego River in Coimbra

This is a structure in reinforced concrete in a multiple frame of five spans designed in 1950 (Figure 12). The deck is 18 m wide. The length between abutments is about 190 m (Soares, 2003). The intermediate supports are pendulous, of free horizontal expansion. The piers are hinged at the base.

2.4. Bridge over the Douro River in Barca d'Alva

The construction of the Almirante Sarmento Rodrigues Bridge over the Douro River, in Barca d'Alva, was completed in 1955

Figure 10. Falsework used to build the Bridge over Sousa River. (a) Elevation. (b) Perspective.

Figure 11. Bridge of Abragão over the Tâmega River. (a) Bridge elevation. (b) Falsework during construction.

Figure 12. Santa Clara Bridge. (a) Bridge elevation. (b) Falsework during construction.

(Figure 13). The bridge is composed by six multiple polygonal concrete arches. Each arch has a span of approximately 40.0 m, and a rise of 7.7 m (ratio 1/5) (Vasconcelos, 2008).

2.5. Bridges over the Cávado and Caldo Rivers

Due to the construction of the Caniçada Dam, it was necessary to replace the old masonry bridges, that were part of national roads EN304 and EN308, by new bridges over the Cávado and Caldo Rivers, with lengths of 206 m and 230 m, respectively (Figure 14) (Soares, 2003).

Edgar Cardoso's solution for these bridges, developed in 1952–53, consisted of a superstructure in reinforced concrete seated on masonry piers and abutments by means of free expanding support devices or rotating fixed supports. The decks are composed by a reinforced concrete slab supported on continuous girders of variable depth with openings in the area of the supports. The decks have, respectively, six and seven intermediate spans of 23 m and two extreme spans of 19 m.

The mobile devices consist of a large roller under each of the deck girders, 1.40 m in diameter and 0.90 m thick. The rough granite masonry piers have a diamond-shaped hollow cross-section of 8 × 4 m on the outside. The highest of these piers is 58 m above the ground, the thickness of which is 0.3 m in the 22.5 m highest zone and 0.40 m in the remaining length. Every 4.5 m a diamond-shaped slab with an elliptical opening in the middle acts as a strap. A side

Figure 13. Bridge over Douro River in Barca de Alva.

hole at the base of the piers enables the free upward and downward movement of the water inside (Soares, 2003).

2.6. S. Fins Bridge

In 1963, Edgar Cardoso designed a bridge over the estuary of Venda Nova Dam (Figure 15), an extremely low concrete arch with a span of approximately 70 m, and a rise of 5 m (ratio 1/14) (IST, 2001).

Figure 16. The former steel bridge of the nineteenth century.

Figure 14. Bridges over the Cávado and Caldo Rivers. (a) Before the closure of the gates of Caniçada Dam. (b) After the closure of the gates of Caniçada Dam.

Figure 15. Bridge over River Rabagão in S. Fins.

2.7. *Mosteirô Bridge over the Douro River*

Due to the construction of Carrapatelo Dam and the consequent rising of the waters of the Douro River, the steel deck of the former nineteenth-century Mosteirô Bridge (Figure 16) was replaced in 1972 by a new and unusual pre-stressed concrete structure. The piers were heightened with reinforced concrete.

The project of this bridge with slightly vaulted arches was carried out by Edgar Cardoso in 1968 (Figure 17). It is an elegant hollow-core rhomboid-type continuous latticed superstructure in pre-stressed concrete, which is supported in the abutments and in two of the three piers of the former nineteenth-century Bridge, which means that the central pier was removed (Figure 18) (Soares, 2003).

The grade line of the deck is a parabola of a 2.5 m rise with a central span measuring 110 m (national record) and two lateral spans with 42 m each. The height of the superstructure varies between the abutments and the piers, and between the piers too, which confers a great elegance on it. This is known as the Edgar Cardoso's favourite bridge. Some say that this shape was inspired in the trajectory of a stone skipping over the water.

3. Arrábida Bridge

For the construction of a new roadway bridge over the Douro River in Porto, Edgar Cardoso studied five different solutions: a reinforced concrete arch, a masonry arch, a suspension bridge, a steel arch and a pre-stressed concrete frame (MOPTC-JAE, 1963). The reinforced concrete bridge had a large central arch with a 270 m theoretical span and 54 m rise, comprising two parallel hollow ribs linked by bracings. The masonry bridge also had two variable thickness masonry ribs. On the plumb lines of the deck piers, the arches had reinforced concrete diaphragms to support the piers and to provide bracing and rigidity for the ribs. The masonry towers on either side of the arch abutments held the lift shafts.

Similarly to the two previous solutions, the steel bridge had one single span straddling the river and both side avenues. The total length of the bridge was 405 m with a 270 m theoretical span and a 54 m rise, consisting of two variable thickness solid-web ribs. The suspension bridge was considerably smaller than usual: its total length was only 513 m, and it had a central span of 220 m and two end spans of 110 m. It was also the most expensive

Figure 17. Mosteirô Bridge elevation.

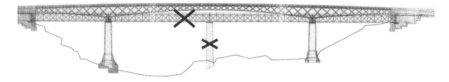

Figure 18. Removal of the central pier.

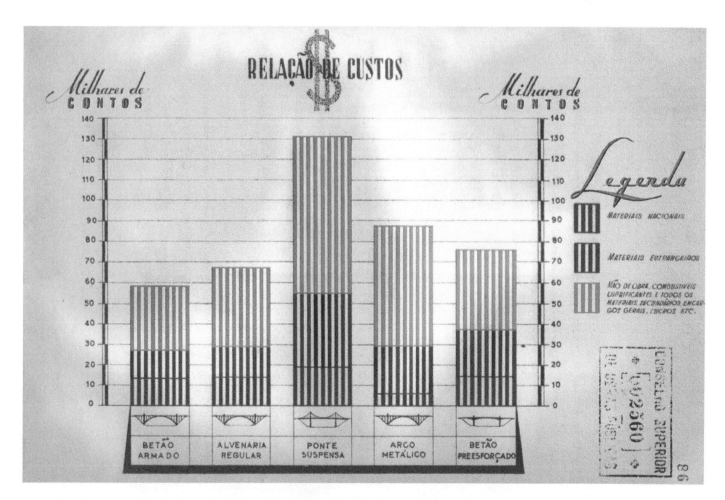

Figure 19. Cost analysis.

solution. The pre-stressed concrete bridge had a cantilever beam with three solid-web spans of varying heights, supported at the ends of the accesses and on two intermediate piers sunk into the banks of the Douro River.

The criteria for evaluating the proposals considered were: the economic factor, the use of national and foreign materials, aesthetics, strength, service life, conservation, construction facilities and risks, and speed of execution (Figure 19). Bearing in mind all these factors, it was decided that all the proposals were technically feasible and, with the exception of the suspension bridge, all were economically viable. However, the most convenient solution

was the reinforced concrete bridge in the 'highest interests of the Nation'. The project is a really detailed piece of art, in terms of descriptive documents, calculations, specifications and drawings.

A revised proposal was delivered in August 1955, and flattering references were made by the rapporteur *the remarkable work consolidating the author's professional merit*s, and by the Ministry of Public Works *the project consisting of a notable piece of work that greatly honours its authors and lends prestige to Portuguese engineering.*

When the bridge was built, the arch span of 270 m was a world record, as it was 6 m longer than the span of the arch of the bridge in Sandö, Sweden. This record span length was later overtaken by the Amizade Bridge between Brazil and Paraguay,

Table 1. World records of concrete arch bridges.

Year	Name	Span (m)	Country
1943	Sandö	264	Sweden
1963	Arrábida	270	Portugal
1964	Amizade	290	Brazil/Paraguay
1964	Gladesville	305	Australia

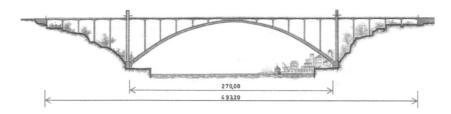

Figure 20. The Arrábida Bridge dimensions.

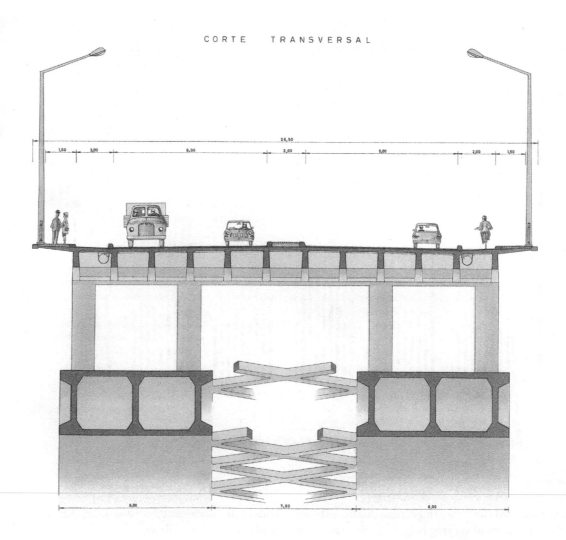

Figure 21. Cross section of the arch and the deck.

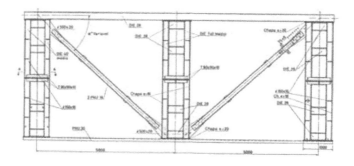

Figure 22. Reduced-size models used for confirming design theoretical calculations. (a) Reduced-size model of the bridge at scale 1/200. (b) Reduced-size model of the deck at scale 1/100.

with more 20 m, and by the Gladesville Bridge in Australia, with more 35 m (Table 1) (Cruz & Cordeiro, 2001).

The arch centreline is polygonal basically coinciding with the anti-funicular of the permanent loads, generating a quasi-uniform state of compression in all the longitudinal fibres (Figure 20). The vertices of the centreline coincide with the points of insertion of the columns (two on each rib of the arch). These columns are hollow and are rigidly linked to the arch and the deck they support. At these points, the double-cell box girder arch sections have diaphragms to prevent the flanges bending.

The final project was slightly altered. The rise was reduced, the spans of the service viaducts were increased, different military loads were considered, and the deck cross-section was now composed of two lanes, two pavements, two bicycle lanes and one central barrier. Following these alterations, the bridge measurements were: (a) deck width: 26.5 m, (b) effective deck width: 25.5 m, and (c) bridge length: 614.6 m (Cardoso, 1955a).

The bridge has twin arches, with a 270 m theoretical span and a 52 m rise; the arches are 4.5 m thick at the abutment and 3.0 m thick at the key, each one forming a double-cell box girder 8.0 m wide, connected by reticulated bracing (Figure 21). This bridge is a good example to illustrate that the *perfection is finally attained not when there is no longer anything to add, but when there is no longer anything to take away* (quote attributed to Antoine de Saint Exupéry).

Figure 23. The steel falsework.

The most modern numerical and experimental techniques were adopted in the bridge design (Cardoso, 1955b). Two reduced-size models of the bridge were made to study the bridge deck. One served to study the slab and the other a segment of deck at the scale of 1/100 (Figure 22(a)). A complete model of the bridge containing almost all its elements was also made at the scale of 1/200 (Figure 22(b)).

The solution of a wooden falsework with foundations in the riverbed was not considered appropriate to Arrábida Bridge

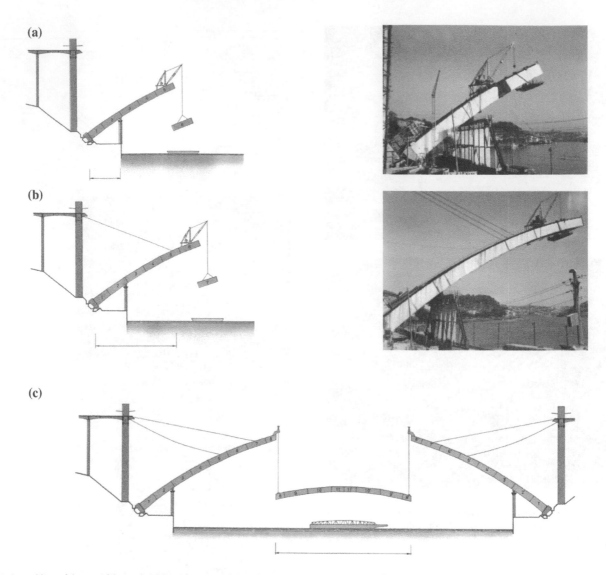

Figure 24. Assembling of the steel falsework. (a) First phase: cantilever (the derrick on the cantilever is lifting a new segment). (b) Second phase: stayed cantilever. (c) Third phase: hoisting the central segment using jacks.

because of the risk of intense floods. In fact, the Douro River has the second-most intense peak flow in Europe, after the Volga River. As an example, between December 2000 and March 2001, five floods were registered with a flow bigger than nine thousand cubic meters per second and average velocities of two meters per second.

One of the greatest challenges of the project was the design of the steel falsework composed by three steel arches, with a 258 m span and a 50 m rise, made with Thomas and Siemens-Martin steel, connected by longitudinal and transverse bracings linking the three ribs (Sobreira, 1963) (Figure 23). It was the first time ever that such a large concrete single-span arch was cast over a deformable steel falsework and that a structure of approximately 2200 tons was moved into place exerting force only on the abutments.

Secheron Portuguesa, S.A. was hired by the general contractor. Under the coordination of Engineer João Cunha de Araújo Sobreira, it built and assembled the falsework in 13 months (between May 1960 and June 1961). The assembly of the falsework was achieved in three phases (Figure 24): in the first phase,

the arch segments adjacent to the abutments were assembled, supported on a provisional concrete frame and hanged from the viaduct decks by means of steel cable stays; in the second phase, the central segment, of about 78 m long and weighing 500 tons, was transported by boat and hoisted with six jacks of 700 tons and 1.0 m stroke, until it rested on the two bank cantilevers that had already been installed; in the third phase, the steel stays were disconnected and the falsework acted as an arch capable of bearing not only its own weight of 2000 tons, but the total weight of the concrete formwork, the dead weight of the concrete arch, and the effects of the wind and earthquake.

The first rib of the arch to be built was the one upstream. Arch segments with varying lengths were successively cast *in situ*. Once all the segments were constructed, the abutment hinges were locked (Figure 25). When the concrete arose the design strength, eight jacks of 700 tons were placed in the key hinge, to introduce a displacement of 10 cm. After this operation, the jacks were moved to the abutments to descend the falsework 40 cm. These works were carried out simultaneously in both banks. The falsework was then moved over a set of rollers to the

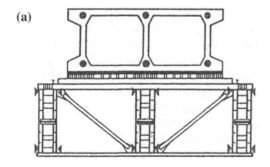

(a)

(b)

Figure 25. Successively cast *in situ* of the concrete arch segments.

second rib, with the action of 100-ton jacks (Figure 26). This was considered the most complex operation of the whole process. After the second arch rib was built, the falsework was moved to a third position to make the bracings between the two ribs (Figure 27). A detailed description of the geotechnical challenges, the short-term and long-term monitoring and rehabilitation works can be found in Cruz and Lopes Cordeiro (2004).

4. S. João Bridge

The increased rail traffic between Porto and Lisbon meant that Maria Pia Bridge (Figure 1) was soon unable to cope with the extra demand made on it. It was a single-track bridge over which trains had to cross at very low speeds (20 km/h). In 1976, a tender for the construction of a new railway bridge in Porto admitted three alternatives: (1) use of the construction falsework used for the Arrábida Bridge in strict compliance with a project to be

(a)

(b)

Figure 26. System used to move the falsework. (a) Side view of the abutment being visible the rollers used to move the falsework. (b) Rollers and guiding rules. (c) Tension system.

prepared by Edgar Cardoso; (2) use of the construction falsework used on Arrábida Bridge following the bidder's solution; and (3) utilise no falsework at all, and an entirely new design proposed by the bidder.

Three bids were submitted: (1) the SEOP-CEPSA consortium proposed Professor Edgar Cardoso's design solution (Figure 28); (2) the SOREFAME-OPCA consortium proposed a solution that made use of the falsework with a design by João Sobreira (Figure 29); (3) the ZAGOPE submitted a proposal with a pre-stressed concrete frame with the piers set in a V-shape, by Engineer Câncio Martins (Figure 30).

The solution presented by Engineer Câncio Martins was finally chosen, but the project suffered a setback when the tender was cancelled in October 1978. In the summer of 1982, the European Investment Bank was contacted to provide the funds. In the meanwhile, Edgar Cardoso signed a contract to draw up the tender process for the bridge and accesses between the train stations of Vila Nova de Gaia and Porto. Actually, Edgar Cardoso produced a base proposal that was very similar to the one previously submitted by Câncio Martins.

On 11 March 1983, the Minister for Housing, Public Works and Transports decided that the design to be put out for tender should be a pre-stressed reinforced concrete multiple-frame bridge. An international tender for prequalification of the companies was launched and about 30 companies replied. Ten of

(a)

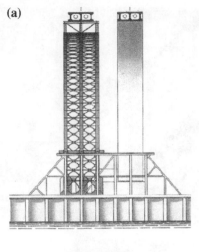

(b)

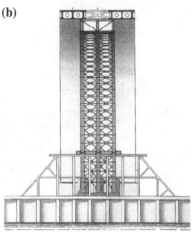

Figure 27. Graphical representation of construction sequence. (a) Falsework under the second arch; (b) Falsework under the bracings.

these companies (two of which were Portuguese) were prequal-ified. The date for opening the bids for this tender had already been set when the tender was again suspended. The object was to find out a solution which, in the eyes of the new government, was more favourable to national and local interests and cheaper to build, an aim to be achieved by harmonising this bridge with the studies for a new road crossing that was being planned for the Freixo area.

Edgar Cardoso suggested a new location closer to the Maria Pia Bridge, about 180 m upstream (Figure 31). This new location was approved by the Board and an addendum was made to the Tender Process, which nevertheless kept the proposal of the piers set in a V-shape although adapted to the new span.

The tender was reopened with this new design, but with the same bidders, and was awarded in May 1984. Edgar Cardoso pre-sented a new solution for a continuous multiple frame on vertical piers (Figure 32). The viaducts have a total length of 528.8 m, the north viaduct 170 m and the south viaduct 358.8 m. The bridge has three spans of 125, 250, and 125 m supported on two large piers set into the riverbed. The central span still holds the world record today for railway bridges of this type (Figure 33).

The piers are supported on enormous reinforced concrete cyl-inders with a diameter of 14 m that are solidly embedded in the bed rock with the aid of 180 micropiles (Soares, 1991). To con-struct the south pier, an octagonal steel platform, supported by 16 piles, was built (Figure 34). Once placed in the river, it would provide access to the steel cofferdam inside it. The cofferdam had 14-m-diameter and was prefabricated in sections, and assembled in rings that were successively lowered using hydraulic jacks from the platform. The lower edge of the steel cofferdam had a double wall forming cells, and it had previously been cut according to the profile of the bedrock on the basis of soundings carried out for this purpose (Figure 35).

Underwater concrete was poured into these cells to seal the contact zone between the cofferdam and the bedrock, and passive anchorages were installed to prevent from being dragged away by river floods. Sealing at the rock interface was completed when divers placed bags of fresh concrete inside the cofferdam. Once the contact zone had been sealed, the water was pumped and the river mud was cleaned. After solving problems of the joints in the granite, the lower part of the base, in which the deepest part was 13.5 m below the average level of the river, was filled with tremie concrete. It was at this level that the micropiles mentioned were placed inside drilled holes.

To avoid the punching shear effect in the reinforced area of the pier base, helicoidal reinforcements were placed on top of the piles (Figure 36). About halfway under the water level, the cylinders became hollow piers with a variable section between the circular crown at the base, which was 12 m across on the outside and 10 m across on the inside, and a rectangular section with rounded edges, measuring 6.535 m on the outside, at the narrowest part at the 46 m level. The pier shafts are hyperboloid in shape, partly truncated by four ruled planes. At 44 m level, the piers cross-section is practically solid, and has only a small hole to install a staircase and an elevator, giving access to the inside of the deck. Edgar Cardoso referred several times that he got the inspiration for this shape when he saw his wife cutting a carrot.

The piers shaft was built by traditional methods using timber formwork, despite the added difficulties due to the complex shape of the formwork and the extremely high density of reinforcements, which in some places were more than 800 kg/m³ (Figure 37). The construction process of the north pier was similar.

The deck, at level 66 m, consists of a trapezoidal section double cell box girder the total height of which varies between 12 m in the sections over the piers and 6 m in the central section and on the transitional piers of the side spans (Figure 38). The three webs vary in thickness from 0.36 to 0.46 m. The thickness of the bottom slab varies from 2.45 m at the first segments and 0.30 m in mid-span and on the viaducts. In addition to longitudinal pre-stressing, the box girders contain 'vertical' pre-stressing. This consists of 1.5-cm-monostrands placed in the three webs to ensure a better performance in terms of cracking. The box girders hold 14 pre-stressed external cables consisting of twenty-four 1.5-cm-chords, 500 m in length, to correct time-dependent effects.

The technique used in building the deck was cast-in-place bal-anced cantilever from each of the two piers, using on-site casting of segments varying in length between 3.0 and 7.5 m. Travellers

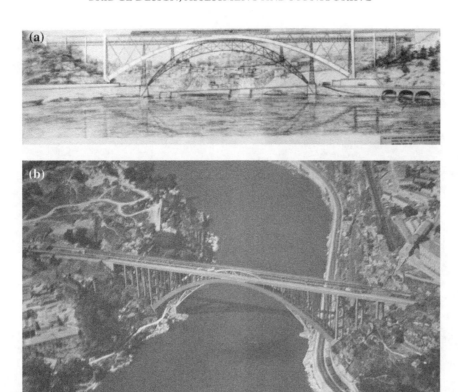

Figure 28. (a) Edgar Cardoso design. (b) Photomontage of the design presented by Eng. João Sobreira, the Maria Pia Bridge being visible, in the second plan (28 m upstream).

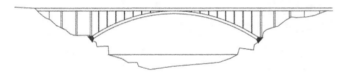

Figure 29. João Sobreira design.

Figure 31. Photomontage of the design presented by Edgar Cardoso in 1982.

Figure 30. Drawing of the proposed design presented by Eng. Câncio Martins.

were used for this purpose and the concrete was pumped up from the base of each pier (Figure 39). In order to cope with the exceptional loads during construction and to allow adjustment operations when casting the concrete in the closure segment of the central deck, temporary steel shoring was erected in the back span (Figure 40); this shoring had hydraulic devices capable of bearing the necessary loads.

The bridge structure, opened on 24 June 1991, was fitted with instrumentation that included survey marks, registers, sensors and measuring instruments to monitor the behaviour of the bridge throughout its existence. Although many were installed and monitored by the LNEC, particularly during the load trials before the bridge opening, additional instrumentation was designed and mounted by Edgar Cardoso himself.

During the project and construction phase, several tests were carried out on prototypes: (a) three pre-stressed concrete beams

Figure 32. New solution presented by Edgar Cardoso in 1984.

Figure 33. View of S. João Bridge.

Figure 34. Octagonal steel fixed platform in the south pier.

Figure 35. Segment of the cofferdam lower ring.

Figure 36. Helicoidal reinforcements placed on the top of the piles at bottom of cofferdam to avoid the punching shear effect.

to determine the rheological characteristics of the concrete; (b) one pre-stressed concrete beam to verify the behaviour of the anchorage zone for an applied force of 6000 kN corresponding to the failure of the pre-stressed steel; (c) one pre-stressed concrete element to verify the behaviour of the deck webs after applying a pre-stressing force of 5000 kN; (d) a segment with the same section as the bridge deck to test the construction processes. In addition to these tests, a two-storey building was constructed near the bridge. The building had two parallel towers built with the same concrete used in the bridge box girder. The static and

Figure 38. Double cell box girder.

Figure 39. Cast-in-place concrete segmental box girder balanced cantilever and form travellers.

Figure 37. Construction of the piers shaft.

Figure 40. Temporary shoring tower.

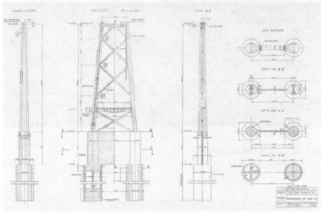

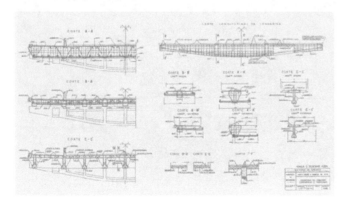

Figure 41. Bridge over Zambezi River.

dynamic behaviour of these towers could be tested, by suddenly releasing the towers, which were previously brought closer with jacks.

5. Bridges in Africa

5.1. *Bridge over Zambezi River in Tete, Mozambique*

Designed in 1962, this is a pre-stressed suspension bridge of funicular cables, without a stiffening beam or expansion joints. Three central spans of 180 m and two end spans of 90 m are supported by four piers with their reinforced, pre-stressed concrete towers (Figure 41). The suspender cables are oblique and form a taut, stable triangular structure. Those cables working together with the inclined suspension cables and the funicular cables make a highly stable structure with very low deflections (Soares, 2003).

Figure 42. Construction of the deck.

Two cables receive the cross girders and the pre-cast slabs in pre-stressed concrete, thus forming the deck (Figure 42). The abutments are of large dimensions to ensure the effective anchorage of the funicular and stiffening cables (Figure 43). Edgar Cardoso carried out the analytical calculations for the bridge and, at the same time, studied the project in various small-scale models as well.

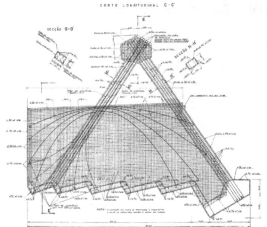

Figure 43. The abutments.

5.2. *Bridge over Save River, in Mozambique*

Built in 1962, this suspension bridge (Figure 44) is similar to the one over the Zambezi River. It is composed of three central spans with 210 m and two lateral spans with 90 m (Soares, 2003).

5.3. *Bridge over Limpopo River in Xai-Xai, Mozambique*

This bridge was built in 1964. It is a cable-stayed bridge with a central span of 120 m and two lateral spans of 35 m (Figure 45). The lateral spans are in pre-stressed concrete, prolonged in 5.0 m to the central span. Those spans connect to steel beams of 37.5 m, which in turn support the central steel beams of 35.0 m. These beams give support to a concrete slab (Soares, 2003).

The concrete towers are articulated in with the piers, and formed by two columns connected by a beam on the top and a diagonal bracing. The piers have indirect foundations that reach almost 80 m deep. The oblique stays are formed by 19 secondary cables and these in turn are divided into 19 strands of high-tensile steel. These cables in a vertical plan that contain the main beams are attached to the pre-stressed beams in the abutments and to the steel beams 72.00 m from the abutments.

5.4. *Bridge over Kwanza River, in Angola*

The bridge over the Kwanza River in Angola, also spelled Cuanza, Kuanza, Quanza or Coanza, is a composite steel-concrete bridge, built between 1970 and 1975. It is a cable-stayed bridge with a total length of 400 m with a central span of 260 m (Figures 46–47) (Rito & Machado, 2002). The towers stand 46 m tall and are supported on indirect foundations that lie at 50 m depth. It is believed that at the time of its construction, the bridge had the largest span of the African continent.

6. A Bridge in China

The Macao-Taipa Bridge is a road bridge linking Macao Peninsular to Taipa (Figure 48). The bridge, built during the time of the Portuguese administration, was the first fixed link

Figure 44. Bridge over Save River.

Figure 45. Bridge of Xai-Xai.

Figure 46. Bridge over Kwanza River.

between the peninsula and the islands of Taipa and Coloane. The population of Macao designates it as 'the old bridge' (Soares, 2003). The construction of the bridge began in June 1970 and was open to traffic in October 1974. The shape of the bridge evokes a dragon (Figure 49).

The total length of the bridge is 2570 m, comprising several precast and moulded *in situ* deck concrete elements from T-beams to box girders. The height of the main structure rises until it passes over the navigation canal with a 73 m span and a minimum gauge of 25 m. At the time of its construction, it was considered the longest continuous bridge in the world. The highest point of the deck is 35 m above the sea level, allowing the passage of ships (Figure 50).

This is a mostly pre-cast bridge (Figures 51–52), to which Edgar Cardoso applied his mastery of special resins and mortars (Soares, 2003). The piers foundations were conceived and made in reinforced concrete piles, driven into the seabed. The required rigidity is achieved by blocks and crowning pieces, also pre-cast, on which the remaining elements of the piers and the deck, also pre-cast, were mounted.

Figure 47. Detail of the steel structure of the deck.

7. Conclusions

Edgar Cardoso was an outstanding bridge designer and a pioneer who broke new ground in the field of experimental analysis. In 1990, his curriculum referred about 500 projects of bridges,

Figure 48. The Macao-Taipa Bridge.

Figure 49. The shape of the bridge evokes a dragon.

Figure 50. The central span.

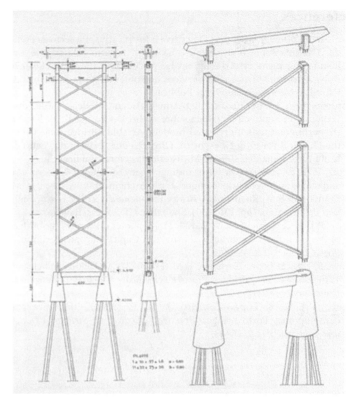

Figure 51. Precasted piers.

Figure 52. Erection of the precast deck.

covering a wide range of typologies and materials (masonry, steel, concrete and composite steel-concrete). He reached world records in big span reinforced concrete arches and in pre-stressed concrete frames. The great diversity of solutions that he conceived represented the own character that he conferred to each work: 'I never make two similar bridges. Each work is always a new solution search'.

Edgar Cardoso passed away on the 5 July 2000, in Lisbon. His legacy constitutes a unique example of work discipline, of full dedication to its profession, of trust in his capacities and intuition, of his desire for excellence. This paper represents a tribute to this audacious engineer and presents a small part of its valuable legacy that deserves a worldwide diffusion and recognition.

Disclosure statement

No potential conflict of interest was reported by the author.

ORCID

Paulo J. S. Cruz http://orcid.org/0000-0003-3170-4505

References

Cardoso, E. (1955a). Memória Descritiva e Justificativa, Projecto da Ponte da Arrábida sobre o Rio Douro, na E. N. 1-Porto [Descriptive and justificative memory, Arrábida Bridge project over the Douro River in E. N. 1-Porto], Direcção dos Serviços de Pontes da Junta Autónoma de Estradas, Ministério das Obras Públicas.

Cardoso, E. (1955b). Ensaios Experimentais em Modelos Reduzidos, Projecto da Ponte da Arrábida sobre o Rio Douro, na E. N. 1-Porto [Experimental tests on reduced models, Arrábida Bridge project over the Douro River in E. N. 1-Porto], Direcção dos Serviços de Pontes da Junta Autónoma de Estradas, Ministério das Obras Públicas.

Cruz, P. J. S. (2014). The outstanding legacy of Edgar Cardoso: An audacious and visionary Portuguese bridge engineer. In A. Chen, D. M. Frangopol, & X. Ruan (Eds.), *Bridge maintenance, safety, management and life extension* (pp. 131–150). Shanghai: CRC Press, Balkema.

Cruz, P. J. S., & Cordeiro, J. M. L. (2001). As pontes do Porto [The bridges of Porto]. Editora Civilização e Porto 2001 Capital Europeia de Cultura, Porto, ISBN 972-26-1890-3.

Cruz, P. J. S., & Lopes Cordeiro, J. M. L. (2003). Audacious and elegant 19th century Porto Bridges. *Practice Periodical on Structural Design and Construction, 8*, 217–225.

Cruz, P. J. S., & Lopes Cordeiro, J. M. L. (2004). Innovative and contemporary Porto Bridges. *Practice Periodical on Structural Design and Construction, 9*, 26–43.

IST – Instituto Superior Técnico. (2001). *Edgar Cardoso – 1913–2000* [Edgar Cardoso – 1913–2000]. Lisboa: Edição Fundação Edgar Cardoso e Departamento de Engenharia Civil e Arquitectura. ISBN 972-957-955718-1-3.

MOPTC-JAE. (1963). *Ponte da Arrábida sobre o Rio Douro e seus Acessos* [Arrábida Bridge over the Douro River and its accesses]. Bertrand (Irmãos), Lda.

Rito, A., & Machado, C. (2002). Inspeção e estudos de reabilitação da Ponte da Barra do Kwanza em Angola [Inspection and rehabilitation studies of Kwanza Bridge in Angola], Conferência Científica e Tecnológica em Engenharia – O Saber do Passado e o Desafio do Futuro. Lisboa: ISEL.

Soares, L. L. (1991). *A Ponte de S. João – Nova ponte ferroviária sobre o rio Douro no Porto* [The S. João Bridge - New railway bridge over the Douro River in Porto]. Ferdouro, A.C.E.

Soares, L. L. (2003). *Edgar Cardoso – Engenheiro Civil* [Edgar Cardoso – Civil engineer]. Porto: FEUP, ISBN 972-752-061-8.

Sobreira, J. C. A. (1963). A montagem do cimbre da Ponte da Arrábida [The assembly of the formwork of the Arrábida Bridge], *Separata da Revista Oficial do Sindicato Nacional dos Construtores Civis*. No. 279.

Vasconcelos, A. (2008). *Pontes dos rios Douro e Tejo* [Bridges of the Douro and Tagus Rivers]. Lisboa: Ingenium Edições Lda, ISBN 978-989-8149-02-2.

Index

Printed and bound by CPI Group (UK) Ltd, Croydon, CR0 4YY

24/10/2024

01778291-0018